Mammals of the
Great Lakes Region

Mammals of the Great Lakes Region

Revised Edition

by Allen Kurta

Drawings by Scott A. Schwemmin
Photographs by James F. Parnell
(unless otherwise noted)

Ann Arbor

THE UNIVERSITY OF MICHIGAN PRESS

1998   1997   1996   1995   4   3   2   1

Library of Congress Cataloging-in-Publication Data

Kurta, Allen, 1952–
    Mammals of the Great Lakes region / by Allen Kurta ; drawings by
    Scott A. Schwemmin ; photographs by James F. Parnell (unless
    otherwise noted). — Rev. ed.
        p.   cm.
    Rev. ed. of: Mammals of the Great Lakes region / William Henry
    Burt. 1957.
    Includes bibliographical references (p.   ) and index.
    ISBN 0-472-09497-1 (acid-free paper). — ISBN 0-472-06497-5 (pbk.
    : acid-free paper)
    1. Mammals—Great Lakes Region.   I. Burt, William Henry, 1903–
    Mammals of the Great Lakes region.   II. Title.
    QL719.A14K87   1994
    599.0977—dc20   94-26780
                                                                CIP

This book is dedicated to
Detroit Cathedral and Michigan State for firm foundations,
Rollin Baker and Tom Kunz for opportunities,
Mom and Pop for quiet support,
Mary for putting up with it all,
and Mark for,
well,
for just being Mark!

Preface

The first edition of *Mammals of the Great Lakes Region,* written by William H. Burt, was published in 1957. Dr. Burt stated that his goals were to provide an up-to-date source of information on mammals living in the Great Lakes basin and to stimulate others to learn more about mammals. To a great extent, he was successful. His book was read by countless amateur and professional naturalists and used by thousands of students, including myself, in classes covering mammalogy, vertebrate natural history, and field biology. The first edition was still in print well into the 1990s, attesting to the book's usefulness.

However, during the 37 years since publication of the first edition, our knowledge of mammals has grown tremendously. My goal in writing the present book is to update the information on mammals, enhance its presentation by adding photographs and new drawings, improve the book's readability, and, at the same time, maintain the simplicity and organization of the original volume. This book is not intended to be a coffee-table book, ponderous tome, detailed listing of relevant literature, or an in-depth treatment of geographic variation. It is intended to serve as a quick reference for teachers, students, naturalists, and professional biologists, and to be a concise guidebook, still small enough to be tucked into a backpack and carried in the field.

Organization

The book begins with a short introduction describing major characteristics of mammals and of the Great Lakes ecosystem. This section outlines physical factors that affect the distribution and abundance of mammals in our area, including surface geology, temperature, snowfall, and vegetation. The Introduction also illustrates the effects, both positive and negative, that European settlers have had on the mammalian fauna.

Most of the book consists of 83 species accounts. Each starts with the animal's common and scientific name and includes information that is usually presented under four headings—Measurements, Description, Natural History, and Suggested References. I include not only native species still living within the Great Lakes drainage but also mammals whose distribution closely approaches our region and introduced species that have become established. Three extirpated species (wolverine, mountain lion, and American bison) are briefly described as well.

In addition to written information, all species accounts have accompanying photographs, and each account contains a small map indicating the species' geographic range in North America and a larger map showing its distribution in the Great Lakes area. I attempted to give a reasonably modern idea of distributions, as opposed to the historical range. Of course, lines drawn on a map only suggest where one is most likely to find an animal, and an individual may wander beyond the areas indicated at any time. Young black bears, for instance, occasionally turn up far south of the suggested range limit when searching for their first home, as do hungry lynx in times of food shortage. Also, one may not find a mammal in the area indicated if appropriate habitat is not suitable; for example, the thirteen-lined ground squirrel is a grassland species not likely to live in dense forests.

Following the species accounts are descriptions of how to capture small mammals and how to prepare specimens for a research or teaching collection, including detailed instructions for making a study skin, fluid preservation, do-it-yourself tanning, and skull cleaning. To help identify unknown specimens, I include two keys that will lead the reader, in a step-by-step process, to the correct identification of an unknown specimen; the first key is based on characteristics of the skull, and the second is based on animals in the flesh. Next are two tables that permit quick comparisons among species. The first table lists body and skull measurements, longevity, litter size, weight at birth, and other useful information. The second table summarizes the number and kinds of teeth present in each species; such dental information often is all that one needs to correctly identify a skull. To improve the readability of the text, I relegated all scientific names of organisms, other than those of mammals of the Great Lakes region, to a list near the end of the book. This listing is followed by a glossary that defines technical terms, the references, and finally an index.

Acknowledgments

In preparing this book, I recorded my own observations and those of my students, although most information was compiled from a variety of published sources. I used primary literature and also frequently referred to *Mammalian Species* (a publication of the American Society of Mammalogists) and various chapters in Chapman and Feldhamer 1982. References that described the mammals of various states were very helpful and included Baker 1983, Gottschang 1981, Hazard 1982, Hoffmeister 1989, Jackson 1961, Merritt 1987, and Mumford and Whitaker 1982. I also consulted books of broader geographic coverage, such as Banfield 1974, Hamilton and Whitaker 1979, Jones and Birney 1988, Nowak 1991, Peterson 1966, and van Zyll de Jong 1983 and 1985. Jones and Manning 1992 and Glass 1981 were helpful in rewriting the keys, and I incorporated ideas found in DeBlase and Martin 1981, Hall 1981, and Nagorsen and Peterson 1980 into the section on specimen preparation. The description of do-it-yourself tanning was adapted from a handout used in Rollin Baker's Mammalogy class at Michigan State University. The common names used in this book generally are those recommended by Jones et al. 1992, and scientific names follow Wilson and Reeder 1993 with a few minor exceptions that reflect common usage or recent discoveries.

North American range maps were based on Chapman and Feldhamer 1982, various *Mammalian Species* accounts, Hamilton and Whitaker 1979, van Zyll de Jong 1983 and 1985, and occasionally Hall 1981. I estimated distributions within the Great Lakes region using published range maps and information obtained from professional mammalogists, government agencies, and museum specimens. Access to museum specimens and/or data was provided by A. Christopher Carmichael and Laura Abraczinskas (Michigan State University Museum), Philip Myers (University of Michigan Museum of Zoology), and Judy Eger and Susan Woodward (Royal Ontario Museum). Additional information on mammal distributions was provided by Charles Long (University of Wisconsin—Stevens Point), Kevin Wallenfang (Wisconsin Department of Natural Resources), Gordon Batcheller and Alan Hicks (New York State Department of Environmental Conservation), and Richard Earle, Tim Reis, and Larry Visser (Michigan Department of Natural Resources).

I would like to thank the University of Michigan Press for allowing me the opportunity to write this book, especially Matt Douglas for convincing me to attempt it and Mary Erwin and Christina Milton for answering innumerable questions and taking care of various problems. I am indebted to A. Christopher Carmichael (Michigan State University), Gordon Kirkland (Shippensburg University), and C. G. van Zyll de Jong (Canadian Museum of Nature) for reading the entire manuscript and making many useful suggestions. Robert Neely and Rollin Baker commented on portions of the manuscript, and Gary Hannan verified botanical names. Philip Myers provided useful tips on pitfall trapping. Joseph Teramino and the students of Zoology 485 (Winter 1994) reviewed the keys, and Paul Ganger established the feasibility of using the wet/dry technique with mice and shrews. Scott Schwemmin spent many hours creating, revising, and again revising the line drawings, and Ron Fraker prepared the range maps. I would also like to thank James Parnell and the other photographers (credited with their photographs) for contributing to this work. Eastern Michigan University provided release time to work on the manuscript through a Faculty Research and Creative Project Fellowship. Finally, I would like to thank my students for putting up with the growls that emanated from my office every time they knocked on the door.

Contents

Mammals of the
Great Lakes Region

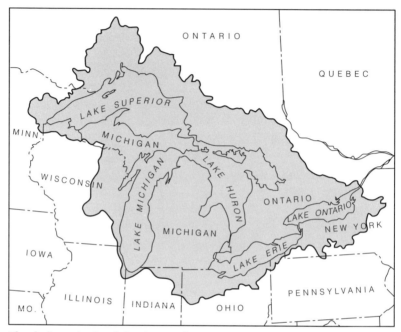

The drainage basin of the Great Lakes

Introduction

What Is a Mammal?

Mammals evolved from an obscure group of reptiles called the Therapsida and first appeared early in the Mesozoic era, about 230 million years ago. Early mammals were probably insectivorous, possibly nocturnal, and generally no larger than modern shrews or mice. They maintained this small size and diversified little, while living in the shadows of dinosaurs and other large reptiles, for over 150 million years.

Early in the Cenozoic era, about 65 million years ago, the mammalian lineage greatly expanded with the sudden appearance of many new species spreading into unfilled niches. This "adaptive radiation" resulted from various factors working in concert. An earlier breakup of supercontinents and other geological events worked to isolate populations and promote speciation. New food resources appeared with the spread of flowering plants and the diversification of such major insect groups as termites, moths, and beetles. In addition, the abrupt disappearance of dinosaurs eliminated major predators and potential competitors. Ultimately, mammals dominated the Cenozoic era, and this time came to be known as the "Age of Mammals."

Today, the class Mammalia contains about 4,600 living species. They range in size from tiny shrews and bats that weigh only 2 g (0.07 oz) to the blue whale, which is the largest *animal* of all time at 160,000 kg (350,000 lb). Modern mammals occupy every major environment from the tropics to the arctic, from deserts to high mountains, and from freshwater ponds to deep oceans. There are mammals that live permanently underground, and still others that spend their lives in trees. There are mammals that run at speeds up to 100 km/hour (60 mi/hour) and others that hop, fly, glide, or swim.

Despite their diversity in size and habits, mammals share a number of distinctive traits. For instance, only mammals have hair. Each hair is an organized collection of cells, once a living part of the skin, but each now dead and filled with the protein keratin. All mammals have some hair, whether it is the thick winter coat of a bear, the embryonic hairs of a whale, or the "whiskers" (vibrissae) on an otherwise naked rodent. For most mammals, hair protects the skin, provides cryptic coloration, and helps regulate body temperature by providing needed insulation.

Another unique trait is a modified skin gland from which the mammals take their name. The mammary glands, of course, secrete milk to nourish developing young and are present in all species, even the egg-laying platypus and spiny anteaters. However, both the number of mammary glands present and the composition of the milk secreted varies greatly among species. Fat content of milk, for example, is 1% in horses, 15% in rabbits, and more than 40% in seals (Oftedal 1984). Mammary number ranges from 2 in most bats to 27 in some marsupials and is roughly correlated with maximum litter size (Gilbert 1986).

Although many vertebrates have teeth, those of mammals generally are more specialized and have much better occlusion, i.e., the upper and lower teeth come together in a more exact fashion. Compare modern reptiles, for example, to mammals. The simple, peglike teeth of a reptile vary little in structure from the front to the back of the mouth, and they are useful in puncturing and holding food, but not much else. In contrast, a mammal can nip small bits of food with its incisors, tear and puncture with its canines, and perform a variety of actions with its cheek teeth (molars and premolars); food may be sliced in a shearing movement, ground between rough edges, or crushed between upper and lower cheek teeth.

The cheek teeth of an omnivore, such as a Virginia opossum or striped skunk, perform all these functions to a certain degree. A strict carnivore, however, like the mountain lion, needs more shearing action to process meaty carcasses, and its cheek teeth are very bladelike. In contrast, a deer consistently chews tough plant fibers; consequently, its teeth have multiple ridges, similar to a file, that grind fibrous food as the teeth slide over one another. Precise occlusion and diverse tooth morphology are keys to mammalian success, helping different species process foods as varied as twigs, leaves, seeds, grass, blood, molluscs, insects, fish, or carrion.

Other identifying characteristics of mammals are largely internal. For example, mammals have an incus, malleus, and stapes (small bones for conducting sound from the eardrum to the inner ear), whereas birds, reptiles, and amphibians have only the stapes. Although both mammals and birds have a four-chambered heart, the mammalian version has a unique left turn to its major artery, the aorta. Mammals are the only vertebrates that have a muscular diaphragm for drawing air into the lungs. In addition, the lungs are partitioned into a vast number of small sacs (alveoli) resulting in an extremely high surface area; this greatly facilitates the movement of oxygen into the blood and permits the high metabolic rate typical of these animals. Mammals can smile, sneer, grimace, or pucker, but other vertebrates lack complex facial muscles and are forever unable to change their expression. Finally, the mammalian brain, particularly the cerebrum, is larger and more complex than in other organisms and endows mammals with a greater capacity for learning and greater overall intelligence. Many other traits unique to mammals are detailed in Vaughan 1986 and Kent 1992.

Mammals and the Great Lakes Ecosystem

The Great Lakes are the largest source of fresh water in the world, and their drainage basin encompasses approximately 770,000 km^2 (297,000 mi^2). This ecosystem extends from northern Minnesota eastward to the headwaters of the St. Lawrence River—a straight-line distance of 1,400 km (850 mi) and a span of 16 degrees of longitude. Its northern boundary reaches beyond Lake Nipigon in Ontario, whereas its southern limit is 1,200 km (725 mi) away in central Ohio. The Great Lakes basin covers almost eight degrees of latitude and is bisected by the 45th parallel. About one-third of the basin is occupied by the Great Lakes themselves. Roughly 70% of the land is within the political boundaries of Ontario and Michigan, and the rest is divided among Minnesota, Wisconsin, Illinois, Indiana, Ohio, Pennsylvania, and New York. More than 20 million humans reside in the Great Lakes region and share its resources with over 75 other species of mammals.

Which mammals occur here, where they live, and how they survive and reproduce is the result of a never-ending interplay between the living and nonliving components of the Great Lakes ecosystem. The purpose of the following sections is to give a brief overview of the geology, hydrology, climate, and vegetation of the Great Lakes basin and to suggest some of the varied ways in which these factors affect the distribution and abundance of mammals. The section concludes with a description of the impact of European settlement on our mammalian fauna.

Bedrock, Glaciers, and Present-day Landscapes

Two features dominate the bedrock of the Great Lakes area—the Canadian Shield in the north and the Michigan Basin to the south (fig. 1). The Shield consists of hard metamorphic and igneous rocks formed during the Precambrian era. These rocks contain extensive mineral reserves (copper, iron, nickel, etc.), and consequently, the Shield has been the site of intense mining activity since the early 1800s. The Michigan Basin, in contrast, consists of much younger sedimentary rocks that were deposited beneath inland seas. A map of the bedrock shows these deposits as a series of concentric circles with the outer rings older than more central ones. However, these are not simply two-dimensional bands of rock. The bedrock of the Michigan Basin is actually a series of saucer-shaped layers nested inside one another. Each ring on the bedrock map represents only the outer rim of a saucer; the middle of each layer is hidden by younger rocks (smaller saucers) that were deposited on top.

During the Pleistocene epoch, glaciers repeatedly scoured, scraped, and generally rearranged exposed bedrock. The hard rocks of the Canadian Shield resisted this erosion to a certain extent. After the ice retreated, a spotty coating of glacial debris was left behind, and many of the ancient rocks were left uncovered. Many of these became the rugged hills of what is now northern Minnesota, northern Wisconsin, northwestern Michigan, and Ontario north of Georgian Bay. The Michigan Basin, in contrast, generally was buried in glacial material; any hills that were left behind were not bedrock but, rather, huge mounds of sand and gravel deposited within the glacier or at its edge. An exception was the Niagara Escarpment, a resistant band of limestone and dolomite of Silurian age that remained above the lakes and glacial overburden and ultimately formed the Door Peninsula of Wisconsin (separating Green Bay from the rest of Lake Michigan), the southeastern portion of Michigan's Upper Peninsula, the islands separating Georgian Bay from Lake Huron, and the Bruce Peninsula of Ontario. The Escarpment continued southward through Ontario as a prominent ridge and eventually formed the basis of Niagara Falls.

The location and type of surface bedrock help explain the distribution

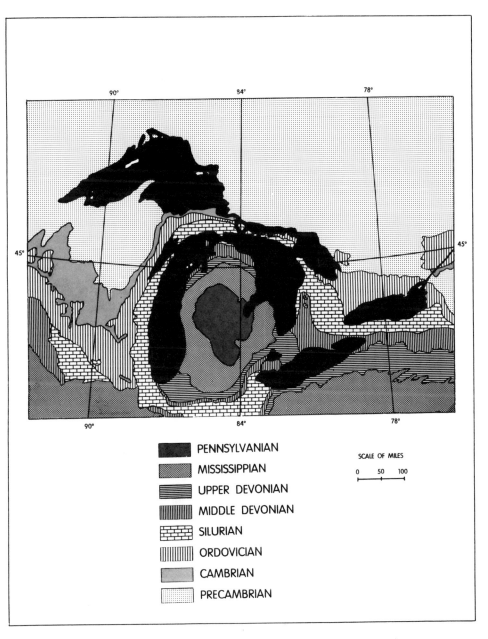

PENNSYLVANIAN
MISSISSIPPIAN
UPPER DEVONIAN
MIDDLE DEVONIAN
SILURIAN
ORDOVICIAN
CAMBRIAN
PRECAMBRIAN

SCALE OF MILES

0   50   100

Fig. 1. Map of the bedrock in the Great Lakes area. (Courtesy of Alan Holman.)

and local abundance of some mammals, especially many bats. Take the eastern pipistrelle as an example. This tiny bat ranges over eastern Ontario, down through New York and Pennsylvania, across southern Ohio and Indiana, and then up into Wisconsin and the western portion of Michigan's Upper Peninsula (see page 86). In other words, the distribution forms a U-shaped bend, seemingly to avoid the Lower Peninsula of Michigan and extreme southern Ontario. This peculiar pattern probably results from the pipistrelle's dependence on caves or mines as winter hibernation sites. In Lower Michigan and southernmost Ontario, all caves are buried under glacial debris, but suitable caves exist near the eastern end of Lake Ontario, in unglaciated areas well south of the Great Lakes drainage, and to a lesser extent in areas to our west. In addition, some of the abandoned mines that dot the Canadian Shield of Ontario, northern Wisconsin, and northwestern Michigan are suitable for hibernation. It appears that the pipistrelle is generally absent from southern Michigan and peninsular Ontario in winter because of the lack of hibernacula, and its absence in summer probably is related to the large migration distance from suitable hibernation sites. The little brown bat and northern bat also require caves (mines) for hibernation, and these mammals are also absent from southern Michigan and extreme southern Ontario in winter; unlike the eastern pipistrelle, however, a few individuals of these species make the long, seasonal migration and become rare summer residents.

The absence of surface bedrock in the southern Great Lakes basin probably affects other mammals as well. For instance, the Allegheny woodrat closely approaches the southeastern boundary of the Great Lakes region. Preferred habitat for this species includes rocky crevices, and the lack of such sites in the southern Great Lakes drainage may partly explain why the woodrat has not spread into our area.

Great Lakes

As the glaciers advanced, major lobes of ice followed ancient stream valleys, broadening and deepening them. Some of these newly formed canyons extended as much as 200 m (700 ft) below modern sea level and contained packed ice well over 1 km (0.6 mi) thick. When the glaciers finally retreated, the largest depressions filled with water and ultimately became the current Great Lakes (Hough 1958).

Many people living outside the Great Lakes region have little appreciation for the immensity of these freshwater "oceans." Few realize, for example, that Lake Superior is as large as the entire state of Maine, or that Lake Michigan is three times larger than Massachusetts. Such huge bodies of water, and their interconnecting rivers, present a barrier to the movement of mammals that is over 2,250 km (1,400 mi) long, stretching eastward from Duluth to Sault Ste. Marie, south to Toledo, and northeast to Montreal.

Small terrestrial mammals have a very difficult time crossing such barriers (Baker 1983), and a quick look at the range maps in this book shows that many distributions end abruptly at a Great Lake or connecting waterway. Arctic shrews, for example, live in Michigan's Upper Peninsula but have not crossed the Straits of Mackinac to the Lower Peninsula. The hairy-tailed mole is abundant on the Ontario side of the St. Marys River, yet it does not exist in adjacent Michigan. The eastern fox squirrel is common in Detroit but not south of the Detroit River, in Windsor. The least shrew inhabits New York state but not the Ontario side of the Niagara River. Even species that occupy both sides of a lake are often represented by different subspecies, indicating a lack of contact between populations (Baker 1983).

Water also is a barrier to mammals dispersing to the hundreds of islands spread throughout the Great Lakes. Individual islands range from small rocks to the 2,183-km^2 (843-mi^2) Isle Royale and the 2,766-km^2 (1,068-mi^2) Manitoulin Island, the largest freshwater island in the world. Some islands occur in archipelagos, whereas others are relatively isolated; many are 20–30 km (12–18 mi) from the nearest mainland. Size, isolation, and, in some cases, younger age contribute to these islands generally harboring fewer mammal species than the mainland. Islands in northern Lake Michigan are a case in point. Ten species of mammals live on South Manitou Island, 13 on North Manitou Island, and 22 on Beaver Island; over 50, however, occur in nearby Lower Michigan (Ozoga and Phillips 1964; Scharf 1973; Scharf and Jorae 1980).

Which species occur on the islands of the Great Lakes is often a matter of chance, but an animal's lifestyle also plays a role (Lomolino 1993 and 1994). For example, bats, with their ability to fly, are often well represented on islands. In contrast, terrestrial hibernating species, such as the jumping mice and the woodchuck, generally are absent. Many mammals colonize islands by walking across the frozen lakes during the winter; hibernators, however, are snug in their underground burrows at that time of year, and dispersal across the ice is not possible for such mammals. Even if a species manages to colonize an island, these populations are prone to extinction. The red squirrel, for example, no longer lives on North Manitou Island even though it was present in the 1940s (Scharf and Jorae 1980).

Little Lakes

Although the five Great Lakes dominate the region, innumerable small lakes and marshes pockmark the countryside from New York to Minnesota. Many in the north were carved out of bedrock or resulted when glacial moraines dammed previously existing low-lying areas. In the south, many formed as "kettle holes." Huge chunks of ice broke off retreating glaciers and became buried under tons of sand and gravel. The ice eventually melted, leaving behind a poorly drained cavity that ultimately became a lake, marsh, or seasonal wetland, depending on its size and the position of the local water table (Dorr and Eschman 1970).

This large number of lakes and wetlands in the Great Lakes region provides ample habitat for a number of semiaquatic species, such as the star-nosed mole, water shrew, beaver, muskrat, and mink. Many other species rely to varying degrees on these smaller lakes for water and/or food. Raccoons, for example, prey heavily on crayfish, and little brown bats feast on mayflies and other insects with aquatic larval stages.

Temperature

Because the Great Lakes region spans eight degrees of latitude, there are significant differences in the amount of solar energy received by northern and southern locations. Northern areas not only average lower temperatures than southern regions but also experience wider swings between summer and winter (Eichenlaub 1979). For example, the average daily maximum temperature in July north of Lake Superior is 23°C (74°F) compared to 30°C (86°F) in northern Ohio; the corresponding temperatures for January are -12°C (10°F) and 2°C (36°F), respectively.

One of the most important ramifications of the north-south temperature difference is its effect on the available growing period for plants (Jones and Birney 1988). Some areas north of Lake Superior have only 60 frost-free days, whereas many southern locations exceed 160 days (Eichenlaub 1979). This difference profoundly affects the seasonal appearance of green vegetation, which in turn affects survivorship, as well as the timing, success, and frequency of reproduction by many resident mammals.

Rain and Snow

Most parts of the Great Lakes basin receive 75–100 cm (30–40 in) of precipitation each year. There is a trend toward greater moisture in the southeast than the northwest and tremendous local variation caused by lake effects and differences in topography (Eichenlaub 1979). This moderate amount of precipitation is a major reason the Great Lakes region at one time was mostly covered by forests rather than the drier grasslands of central North America. However, the small differences in annual moisture within the basin have little, if any, direct impact on mammal distributions.

Snow is a different matter. Annual snowfall across the Great Lakes drainage differs greatly with latitude and position in respect to the lakes (Eichenlaub 1979). Prevailing winds in winter are from the north and west; consequently, the southern and eastern shores of the lakes receive the most snowfall. Annual amounts of 500 cm (200 in) or more occur in such restricted locations as the Keweenaw Peninsula of Michigan, Blue Mountain region of Ontario, or Tug Hill Plateau of New York. In addition to lake effects, there is the expected trend of more snowfall in the colder north. Average yearly snowfall is more than 250 cm (100 in) north of Lake Superior but well below 100 cm (40 in) in northern Indiana and Ohio. Cooler temperatures in the north also mean that snow remains for longer periods, continually covering the ground for 6–7 months in many areas (Eichenlaub 1979; Strong 1989).

Mouse-sized mammals are not adversely affected by a continual cover of deep snow and actually benefit from it. Most simply continue their normal activities in the subnivean environment, travelling to and fro through snowy tunnels. The snow protects them from the eyes of predators and insulates against colder air above. For instance, a caribou standing in a spruce forest may experience an air temperature of -15°C (5°F), but air surrounding a deer mouse, 60 cm (2 ft) beneath the snow, is at a balmy 0°C (32°F). The *lack* of snow in late autumn or early winter actually increases mortality in small mammals, especially when coupled with unusually low temperatures (Merritt and Merritt 1978).

In reality, deep and prolonged snow cover is hardest on medium-to-large mammals because it interferes with mobility and makes food detection, prey capture, or predator escape more difficult. Simms (1979a), for example, speculates that snow is one factor limiting the northward distribution of the long-tailed weasel. Presumably, this carnivore is too large to follow prey through subnivean tunnels, and the weasel moves awkwardly over the surface of deep snow because it lacks the broad feet of a wolverine or lynx. The smaller ermine, in contrast, travels easily beneath snow, and its range continues up to Hudson Bay. As another example, Gates, Clarke, and Harris (1983) point out that the northern limit for white-tailed deer fluctuates from year to year as a function of temperature and snow cover. The deer handle cold winters well and snowy winters well, but a cold and snowy winter is devastating; low temperatures increase the demand for energy, yet snow makes food more difficult to obtain. Bobcats face a similar problem because they have difficulty chasing prey in deep, powdery snow (McCord and Cardoza 1982).

Vegetation, Plant Communities, and Biomes

Ecologists typically divide the earth into a handful of biomes—each a broadly defined community that results from complex interactions among precipitation, temperature, soil, and vegetation. Although actual vegetation varies greatly on a local scale (Albert, Denton, and Barnes 1986; Chapman and Putnam 1973; Strong 1989), the biome concept allows one to make some useful generalizations. Two such communities dominate the Great Lakes region—the Coniferous Forest Biome and the Deciduous Forest Biome (fig. 2). The boundary between them is indistinct, and a broad transition occurs through central Wisconsin, south of Green Bay; across Michigan's Lower Peninsula, from Muskegon to Tawas City; and from the southern end of Georgian Bay to the northern shore of Lake Ontario.

The area to the north is part of the Coniferous Forest Biome, and it is characterized by moderate rainfall, cool temperatures, and soils that are often sandy, somewhat acidic, and low in organic matter (Clayton et al. 1977; Veatch 1953). The climax forest north of Lake Superior contains primarily spruce and fir interspersed with broad-leaved trees, such as aspen or birch. South of Lake Superior, but north of the transition, the forest is more varied. There are extensive tracts of pine as well as large stands of northern hardwoods (maple, beech, and birch) mixed with evergreen trees, especially fir and hemlock (Braun 1964). A number of

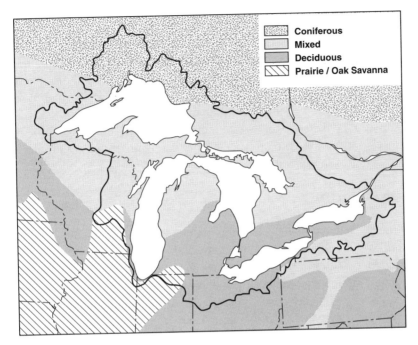

Fig. 2. Map of major plant communities in the Great Lakes basin

mammals live primarily in the Coniferous Forest Biome and rarely venture into the Deciduous Forest Biome; these include the Arctic shrew, water shrew, northern flying squirrel, southern red-backed vole, marten, fisher, lynx, and moose, among others.

The Deciduous Forest Biome, in contrast, is characterized by warmer temperatures and soils that contain more clay and more organic matter and are less acidic (Clayton et al. 1977; Veatch 1953). Depending on location, mature forests are dominated by beech-maple, elm-ash, or oak-hickory associations (Braun 1964). Within the Great Lakes region, the least shrew, eastern mole, and woodland vole are a few of the species largely restricted to this biome.

A third major community, the Grassland Biome, closely approaches, and occasionally enters, the Great Lakes drainage along its western and southwestern boundaries (Braun 1964). Unlike the vast prairies of central North America, the Great Lakes version of this community originally was fragmented and savannalike with bur oak trees punctuating the grassy lowlands and stands of white and black oak dominating the neighboring sandy ridges. The deep, rich soils of these grasslands were quickly converted to farmland by European settlers, and little of the original prairies is left in our area. Although the Grassland Biome covers only a minute portion of the Great Lakes drainage and has been drastically altered by humans, this biome contributes to the diversity of mammals in our area. The plains pocket gopher, western harvest mouse, Franklin's ground squirrel, and prairie vole are species that barely enter our region along with their grassland habitat.

Human Impact

Although Native Americans undoubtedly affected local mammal communities, European settlers profoundly influenced the distribution and abundance of many mammal species in a relatively short time. Large predators competed with these immigrants for edible wildlife and domestic stock and suffered greatly because of it; the mountain lion, for example, was extirpated from the Great Lakes area, and the gray wolf and Canadian lynx were driven northward into areas less inhabited by humans. The ever-growing European population persistently hunted elk, moose, caribou, and bison until these species either disappeared or retreated northward as well. Overtrapping eliminated the beaver, river otter, fisher, and marten from many parts of their former range, and habitat degradation contributed to the decline of these and other species.

However, human impact was not always negative. The eastern fox squirrel, for instance, dramatically expanded its range during the last two centuries. This squirrel preferred a mosaic of forests and grassy openings, a habitat that originally was common only along the southwestern boundary of the Great Lakes basin. The vast forests that originally cloaked the southern Great Lakes region were not to its liking, but extensive logging and land clearing eventually created a plethora of desirable "edge" habitats. The fox squirrel rapidly invaded these areas and today lives 400 km (250 mi) farther north than it did in 1800. The white-tailed deer, thirteen-lined ground squirrel, and coyote also became more abundant after settlers and loggers decimated the once-great forests.

Bats benefitted from human interference as well. Mining activities in the Canadian Shield created hibernation sites that did not exist prior to 1800 and allowed some bats, such as the eastern pipistrelle, to extend its range. At least two species, the little brown bat and big brown bat, abandoned hollow trees as summer roosts and came to rely almost exclusively on barn rafters, house attics, and similar structures; surely these mammals became more abundant after the European invasion.

Humans also affected the mammalian fauna by intentional and unintentional introductions. White-tailed jackrabbits, as an example, were released into Wisconsin as a game animal, and the descendants of European hares that escaped captivity eventually spread throughout southeastern Ontario. Although elk originally lived in the Great Lakes region, those present today trace their ancestry to animals imported from western North America. Finally, the commensal house mouse and Norway rat hitchhiked from Europe to North America and colonized the Great Lakes area along with their human benefactors.

Species Accounts

Marsupials
Infraclass Metatheria

Marsupials are a fascinating group of mammals that was once widely distributed across North America, South America, Europe, Australia, and even Antarctica. Today, most of the remaining 260 species are restricted to Australia and its adjacent islands and to Central and South America, although one species ranges northward into the United States and Canada. Despite the small number of living species, present-day marsupials are very diverse. Most people are aware of kangaroos and wallabies, but there are also marsupials that strongly resemble moles, mice, flying squirrels, ground squirrels, rabbits, weasels, and a number of other more familiar mammals. Living marsupials vary in size from only a few grams (less than an ounce) to almost 100 kg (220 lb).

The most unique aspects of marsupial biology concern reproduction. Gestation periods are very short compared to the so-called placental or eutherian mammals, and even the largest marsupials are born after only 10–40 days of uterine development. At birth, the young are still nothing more than poorly developed embryos lacking fully formed hearts, lungs, and kidneys; they cannot see or hear and are naked and exceedingly small. A newborn red kangaroo, for example, weighs less than 1 g (0.04 oz) even though its mother's mass approaches 30 kg (65 lb).

At birth, marsupial young crawl unaided from the vagina and upward through the mother's fur searching for a nipple. Nipples are usually hidden in a pouch (marsupium), but this protective structure is lacking in some marsupials, such as the South American mouse opossum; pouches open either to the front, as in the kangaroos, or to the rear, as in bandicoots. Once the newborn locates a nipple, the youngster stays firmly

17

attached to it for several weeks as development proceeds. The young marsupial eventually makes brief forays from the pouch, and these become longer and longer in duration, until the offspring is totally independent. Although gestation is brief, total time between conception and weaning is generally longer than in eutherian mammals, indicating that marsupials actually have slower rates of development.

Marsupials differ from eutherian mammals in a number of other structural and physiological characteristics. For example, the vagina of marsupials is branched; the penis of the male often is correspondingly forked and located in back of, rather than in front of, the scrotum. In most marsupials, the placenta is made from fetal membranes called the chorion and the yolk sac, instead of the chorion and allantois that are used by eutherians. Many marsupials have more than 44 teeth, whereas most eutherians have fewer than that number. In marsupials, the angular process of the lower jaw is inflected (points inward), and the jugal bone is quite large, reaching back as far as the jaw joint. Marsupials generally have smaller brains than other mammals of comparable body mass. In addition, the body temperature, metabolic rate, and heart rate of marsupials are usually lower than those of eutherians. Despite their multiple differences, marsupials are in no way inferior to eutherian mammals; the two groups have simply evolved different solutions to similar problems.

New World Opossums
Order Didelphimorphia
Family Didelphidae

The marsupial order Didelphimorphia contains a single family, the Didelphidae. Members of this New World family range from mouse-sized creatures up to the 5.5-kg (12-lb) Virginia opossum. These marsupials usually have short, woolly fur and a long, scaly, prehensile tail. Legs are rather short, and both front and rear feet have five well-formed digits. All toes are clawed, except the hallux or big toe; this opposable digit is functionally equivalent to the human thumb, allowing the animal to grasp and manipulate objects with its rear feet. Most didelphids are omnivorous and arboreal, although a few species are primarily ground dwelling. One unusual didelphid, the water opossum, leads a semiaquatic existence and comes equipped with webbed feet, water-repellent fur, and a waterproof pouch. The Didelphidae has a rich fossil history dating back to the late Cretaceous period, and today there are 69 living species distributed among 15 genera. Although this family is primarily tropical in distribution, one hardy species, the Virginia opossum, has invaded temperate areas as far north as the Great Lakes region.

Virginia Opossum *Didelphis virginiana*

Measurements: Total length: 650–850 mm (26–33 in); tail length: 250–380 mm (10–15 in); hindfoot length: 50–80 mm (2–3 in); ear height: 40–60 mm (1.6–2.5 in); weight: 2.0–5.5 kg (4.4–12.0 lb).

Description: A Virginia opossum is about the size of a large house cat. Its cone-shaped head has an elongate snout, pinkish nose, and small eyes. The ears are mostly black, rounded, rather thin, and hairless, and

Virginia opossum

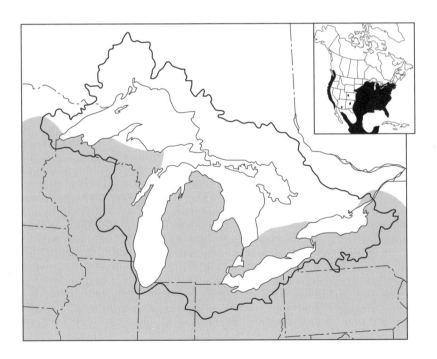

the long, naked tail is light-colored and prehensile. Both ears and tails often show damage from frostbite. This mammal has an overall grayish-white, dishevelled appearance, resulting from white underfur and outer hairs that are all white or white tipped with black. As in other didelphids, all toes are clawed, except the opposable hallux. Females possess a forward-opening pouch. The skull has a small rounded braincase, prominent sagittal crest, and more teeth (50) than any other mammal in our region (fig. 3).

Natural History: This marsupial is really a southern species. It probably was absent from the Great Lakes basin prior to 1800, but it slowly spread into our area during the past two centuries and is expanding its range northward even today. Although the Virginia opossum favors deciduous woods near a stream or lake, it also frequents semiopen country, brushy fencelines, drainage ditches, and swamp borders. It is an adaptable species that flourishes in urban parks as well as rural areas.

Although the opossum most often rests in an abandoned woodchuck burrow, it occasionally finds shelter in a hollow tree, brush pile, culvert, or building. This marsupial lines its den with dry grass or leaves that it carries home using its mouth and prehensile tail. A Virginia opossum leads a solitary existence, but it does not defend a territory and does not have a well-defined home range. Instead it is a wanderer, switching dens every two days or so. It moves an average of 300 m (980 ft) between sites and travels anywhere from 10 to 600 m (30 to 2,000 ft) away from the den in search of food each day. The opossum generally forages at night, relying mostly on the senses of smell and hearing to locate edible items.

When confronted by an enemy, an opossum often feigns death, i.e., it "plays possum." The animal falls on its side with eyes open, teeth barred, and mouth slightly open, drooling heavily; the opossum does not respond when touched and remains in this catatonic state for a few minutes up to six hours. The value of this peculiar behavior is open to question, although it may startle a naive predator and allow the opossum to escape eventually.

This mammal is a true omnivore. Common animal foods include earthworms, insects, and bird eggs; small mammals, nestling birds, amphibians, and even snakes are eaten when encountered. Rural opossums readily scavenge road kill and other carrion, whereas garbage and uneaten dog food are popular with urban dwellers. When available, fruits, nuts, seeds, and mushrooms supplement the diet.

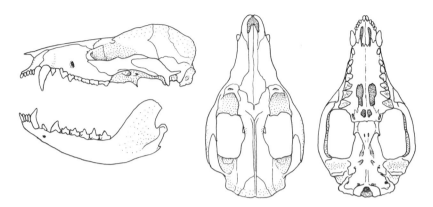

Fig. 3. Skull of a Virginia opossum, family Didelphidae

The Virginia opossum mates in mid- to late winter, and young are born 13 days later, usually in March. In the southern Great Lakes region, a few females rear a second litter in late June or July. Each newborn is poorly developed and weighs only 0.15 g (0.005 oz); it would take 20 newborn opossums to equal the weight of a penny. Although litters occasionally contain as many as 17–21 young, 7–9 offspring are more typical. Young opossums suckle continuously for two months before venturing out of the pouch; they then remain with their mother for an additional month, often riding on her back as she forages. Young females become sexually mature at about 8 months of age and first breed in the year following their birth.

Average life expectancy is less than two years. Harsh winter weather undoubtedly causes some mortality, and thousands are killed each year by automobiles. In addition, many fall prey to hawks, owls, dogs, and foxes, as well as the occasional human hunter. Snakes are less of a problem because the Virginia opossum is immune to rattlesnake venom.

Suggested References: Gardner 1982; Hossler, McAninch, and Harder 1994; McManus 1974.

Placental or Eutherian Mammals Infraclass Eutheria

There are three major reproductive strategies among mammals. First, the monotremes retain the ancient reptilian pattern of egg laying. Second, the marsupials rely on a brief period of intrauterine development, followed by a prolonged interval of postnatal growth, usually within the confines of a maternal pouch. The third strategy, found in the eutherian mammals, includes the development of a more complex placenta that allows a growing fetus to stay within the uterus for longer periods than a typical marsupial. Although gestation length is as short as 18 days in some eutherians (shrews), it may last as long as two years in others (elephants).

Modern eutherians occur naturally on all continents and include over 90% of living mammal species. Eutherians have adopted a tremendous array of life-styles, including some that have eluded the monotremes and marsupials. For example, only placental mammals have evolved totally aquatic forms (whales), and only placentals have developed powered flight (bats).

Shrews, Moles, and Their Allies
Order Insectivora

Insectivores are an ancient group of eutherian mammals that dates back to the Cretaceous period, almost 130 million years ago. Today, these mammals are widespread on most continents, although they are missing from Australia and much of South America. Modern insectivores live in a variety of habitats, including sand dunes, deserts, tropical jungles, and high mountains; most are terrestrial or fossorial, but a few species, such as the giant otter shrew of Africa, are semiaquatic. Although fewer than 450 species exist, the order contains many intriguing forms. The South African golden mole, for example, has iridescent fur, and the Haitian solenodon and North American short-tailed shrew are unusual among mammals in that they use venom to subdue prey.

The insectivores are a diverse group, and there is no single characteristic that readily distinguishes them from other mammals. Insectivores are generally small bodied, ranging in mass from only 2 g (0.07 oz) to about 4 kg (9 lb), the size of a large rabbit. Members of this order typically have five clawed toes on each foot and walk on the soles of their feet (plantigrade locomotion). Eyes are small and beady, and the elongate snout extends forward well beyond the jaw. The fur is usually soft, short, and dense, but a few groups, such as hedgehogs and tenrecs, modify some hairs into hard protective spines.

Although most insectivores eat insects, many supplement these with a variety of other invertebrate and vertebrate prey. The giant otter shrew, for instance, is fond of freshwater crabs, and the short-tailed shrew readily eats voles and mice. Most species, including all those in the Great Lakes area, have cheek teeth with W-shaped ridges (see fig. 4). When the animal chews, these ridges slide past each other in a scissorslike action, slicing through the tough body parts of their prey.

24

Shrews
Family Soricidae

About three-quarters of all insectivores belong to this family, including some of the smallest mammals in the world. The Old World species *Suncus etruscus* and the New World *Sorex hoyi* are called pygmy shrews. Some adults of these species weigh a mere 2–3 g (0.07–0.11 oz) and measure only 4 cm (1.6 in) from nose to tip of tail; among mammals, only the bumblebee bat of Thailand may be smaller.

Shrews have small functional eyes, yet these animals do not depend on vision for finding prey or moving about their environment. They gather much information through the sense of touch using their sensitive snout and large vibrissae. Furthermore, hearing is well developed. These insectivores are attracted to sounds made by potential prey, and some shrews, such as our short-tailed shrew, are echolocaters. They actually detect obstacles, and perhaps prey, by emitting high-frequency sounds and listening for the returning echoes. The sense of smell also is acute in shrews; they use it to locate buried food, such as insect pupae, and to identify prey encountered in underground tunnels.

Many shrews have well-developed scent glands to go along with their heightened sense of smell. The glands typically are located on the flank and exude a musky odor used to attract mates or mark territorial boundaries. Moreover, these secretions provide a degree of protection for their owner; many predators apparently consider the odors revolting and hesitate to attack or eat a shrew because of them.

Although some shrews in the Great Lakes region appear mouse sized, one can readily distinguish them from rodents and other small mammals by a number of characteristics. Compared to rodents, shrews have a longer and more conical snout, smaller eyes, and shorter fur. The external ear of shrews is minute, and the ear opening is hidden by fur. In addition, shrews have five well-formed toes on their forefeet compared to only four in small rodents. Moles are somewhat similar to shrews, but moles generally are larger and have huge, paddle-shaped forefeet.

The skulls of our shrews also are distinctive (fig. 4). For example, the tips of most shrew teeth are reddish brown; an iron compound, deposited during growth, hardens the teeth and causes this coloration. Similarly sized skulls of bats, rodents, and moles lack such pigmentation. In addi-

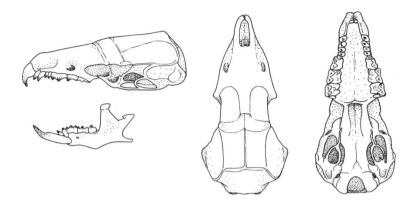

Fig. 4. Skull of a northern short-tailed shrew, family Soricidae

tion, the skull of a shrew is missing the zygomatic arch that is present on the skulls of all other mammals in our area.

Arctic Shrew   *Sorex arcticus*

Measurements: Total length: 105–125 mm (4.1–4.9 in); tail length: 37–46 mm (1.5–1.8 in); hindfoot length: 12–15 mm (0.5–0.6 in); weight: 7–13 g (0.2–0.5 oz).

Description: The Arctic shrew is larger than most shrews in the Great Lakes region, and its body fur has a distinct tricolored pattern. It is black or dark brown across the head and back, light brown along the sides, and grayish brown on the belly. The tricolored appearance is most obvious in winter-caught specimens but poorly developed in juveniles. The tail is indistinctly bicolored, grading slowly from its dark dorsal surface to a slightly lighter brown below.

Natural History: The Arctic shrew is a northern species that reaches the southern limit of its distribution in the Great Lakes region. Even though it is common in some areas, biologists know little of its habits and lifestyle compared to most of our other shrews. This species occasionally lives in fairly dry fields and forest openings, but it prefers moist areas adjacent to a lake, bog, swamp, or ditch. Although it favors damp areas, a very high water table leads to a decline in Arctic shrews and an in-

Water shrew (top), Arctic shrew (middle), and pygmy shrew (bottom). Range map is for Arctic shrew. (Photo by R. E. Wrigley, Manitoba Museum of Man and Nature.)

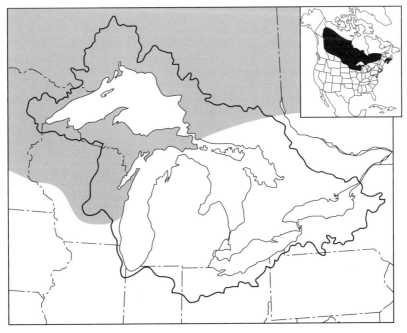

Arctic shrew

crease in the masked shrew population. Densities of 8–12/ha (3–5/acre) are typical for the Arctic shrew.

The Arctic shrew is active throughout the day and night and probably solitary in habits. In captivity, it is quite docile and seldom bites. It is extremely energetic, always running and never walking. When resting, this shrew takes on a slightly curled posture, lying on its side or belly with the head tucked underneath. The Arctic shrew is mostly silent, though it occasionally emits a rapid chatter while dashing about. It grooms by wiping the forefeet along the mouth and apparently drinks by lapping water with its tongue.

In the wild, an Arctic shrew feeds mostly on insects, such as sawfly larvae and adult grasshoppers. It generally searches for prey along the ground but may climb vegetation up to 30 cm (1 ft) while foraging. In captivity, this shrew readily consumes dead voles, yet it does not attack live ones. Other foods eaten in captivity include fly pupae and mealworms.

Mating begins in late winter, and young appear as early as the end of April; the exact gestation period is not known, but it probably is 20–23 days. Although litter size varies from 4 to 10, it averages about 7. Breeding continues into September, and some females produce two or more litters each year. A few females reproduce during the year of their birth, but most wait until the following spring. Maximum longevity reported for an Arctic shrew in the wild is just 18 months. Owls are the only known predator.

Suggested References: Baird, Timm, and Nordquist 1983; Buckner 1970; Clough 1963.

Masked Shrew   *Sorex cinereus*

Measurements: Total length: 75–100 mm (3–4 in); tail length: 31–45 mm (1.2–1.8 in); hindfoot length: 10–13 mm (0.4–0.5 in); weight: 3.5–5.5 g (0.1–0.2 oz).

Description: Along the back and sides, a masked shrew appears mostly brown with a hint of gray that is particularly evident in strong light. Ventral hairs are grayish at the tip and black at the base; hence, the chest and belly appear somewhat mottled with gray and black. The tail is lightly furred and slightly darker above than below.

Masked shrew

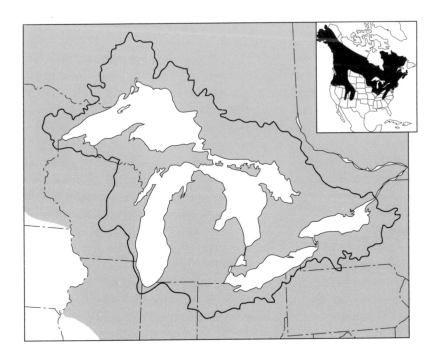

Within the Great Lakes basin, the masked shrew is often indistinguishable in size and color from the pygmy shrew, and the best way to differentiate between these two is by examining the unicuspids. These are small teeth, simple in structure, located behind the well-formed first incisor and in front of the first large cheek tooth. There are five unicuspids in most of our shrews; however, not all are obvious when one looks at the toothrow from the side. There are, for example, only four unicuspids readily visible in the *side* view of the short-tailed shrew skull shown in figure 4. A masked shrew also has four obvious unicuspids when viewed from the side, in contrast to a pygmy shrew that has only three (also see fig. 30 in the Key to Skulls).

Natural History: The masked shrew is widely distributed across northern North America and is one of the most common mammals in the Great Lakes area. Preferred habitat is moist woodlands containing abundant plant cover, thick leaf litter, and decaying logs. However, the masked shrew tolerates a wide range of conditions, and one often finds it in overgrown fields, alder thickets, cedar swamps, weedy fencerows, grassy marshes, and even sphagnum bogs. It is least common in dry upland areas and exposed fields. Population density typically ranges from 7 to 28/ha (3 to 11/acre) but occasionally surpasses 200/ha (80/acre) under exceptional conditions.

The masked shrew is a restless loner that is active at any time of day or night, in either winter or summer. It frequently travels next to fallen logs or uses obscure paths through the leaf litter, but it also scampers through mole tunnels and readily follows grassy runways built by voles. A masked shrew rests periodically in a nest fashioned from leaves and grass and hidden inside a half-rotted log or beneath a rock. The nest measures 7–10 cm (3–4 in) across on the outside, but the diameter of the inner chamber is less than 1.5 cm (0.6 in)—just enough room for a single shrew. Like many species of shrew, the masked shrew adapts poorly to captivity and often dies soon after capture, apparently from shock or fright.

In the wild, a masked shrew consumes just about any type of terrestrial invertebrate that it stumbles across. The most common foods are caterpillars and grubs, but adult insects, especially beetles and crickets, are frequently on the menu as well. Other important prey include slugs, snails, spiders, and, to a lesser extent, earthworms and centipedes. This tiny mammal rarely, if ever, attacks vertebrates, although biologists occasionally find salamander flesh or mouse fur inside the stomach of a masked shrew. Plant material generally makes up less than 1% of the diet, and most of this may be accidentally ingested while the shrew consumes its prey.

The masked shrew combines a long mating season (March to September) with an extremely short gestation (18 days); consequently, adult females produce up to three litters each year. Litter size varies from 2 to 10, but most females give birth to 5–7 offspring that weigh 0.25 g each (0.01 oz). At birth, a masked shrew is better developed than a Virginia opossum, yet poorly developed compared to many other mammals. A newborn is naked, blind, deaf, and toothless; its toes are still fused together, and the stomach, liver, and intestines are readily visible through the thin abdominal wall. Growth is rapid, however, and weaning takes place when the youngsters are about three weeks old. Individuals reach sexual maturity after only two months.

Lifespan typically does not exceed 18 months. Mammalian predators include red fox, coyote, weasels, and other shrews. Snakes, owls, and even large frogs also prey upon the masked shrew.

Suggested References: Forsyth 1976; French 1984; Ryan 1986.

Long-tailed or Rock Shrew   *Sorex dispar*

Measurements: Total length: 110–135 mm (4.3–5.3 in); tail length: 50–64 mm (2.0–2.5 in); hindfoot length: 14–15 mm (0.6 in); weight: 4–6 g (0.1–0.2 oz).

Description: Throughout the year, a long-tailed shrew has a slate-gray pelage that is essentially the same color on dorsal and ventral surfaces. A large body, unusually long snout, lengthy tail, and a restricted geographic range should readily distinguish a long-tailed shrew from most other shrews in our area. One might confuse this species with a smoky shrew, particularly in winter when the latter also takes on a slate-gray appearance; however, the smoky shrew often has belly hairs that are slightly lighter than those on the back. Furthermore, tail length in a long-tailed shrew averages 85% of head and body length, compared to only 65% in a smoky shrew. The tail of a long-tailed shrew also tends to be thicker and not as distinctly bicolored as that of a smoky shrew. The

Long-tailed shrew. (Photo from the Roger W. Barbour. Collection, Morehead State University.)

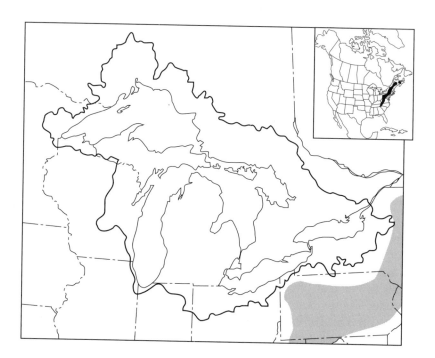

moderately furred tail of a young long-tailed shrew becomes naked as the animal ages, a pattern often seen in other members of the genus *Sorex*.

Natural History: This mammal occupies the Appalachian Mountains from North Carolina to Maine, and its distribution most closely approaches the Great Lakes drainage in New York and Pennsylvania. The aptly named long-tailed shrew lives in rocky outcrops, among moss-covered boulders and stones, within cool mountain forests. Here, its slender snout allows it to gather food from narrow cracks among the stones, and its long, thick tail serves as a balancing organ when the animal scampers from boulder to boulder. In addition, the long-tailed shrew occasionally lives along mountain streams in areas frequented by water shrews. Other small mammals found in long-tailed shrew habitat include southern red-backed voles, rock voles, deer mice, and smoky shrews.

Biologists are not small enough to enter the nooks and crannies occupied by the long-tailed shrew, and we know few details of its ecology, reproductive habits, or longevity. The diet consists of centipedes, spiders, and insects, such as grasshoppers, flies, and beetles. The mating season apparently lasts from April into August, and litter size is 2–5.

Suggested Reference: Kirkland 1981.

Smoky Shrew   *Sorex fumeus*

Measurements: Total length: 110–126 mm (4.3–5.0 in); tail length: 42–52 mm (1.7–2.0 in); hindfoot length: 12–15 mm (0.5–0.6 in); weight: 6–11 g (0.2–0.4 oz).

Description: Hairs along the back are slate gray to black in winter, whereas they have a dull brown color in summer; regardless of season, belly hairs are similar in color or just slightly lighter than those above. One of the smoky shrew's most distinctive traits is a bicolored tail that appears dark on top and abruptly changes to a yellowish tan underneath. This species is similar in size to the widespread Arctic and water shrews; however, a smoky shrew lacks the tricolored appearance of an Arctic shrew and the hairy fringe found on the feet of a water shrew.

Natural History: The smoky shrew lives along the Appalachians and in the eastern part of the Great Lakes basin. This insectivore occasionally

Smoky shrew

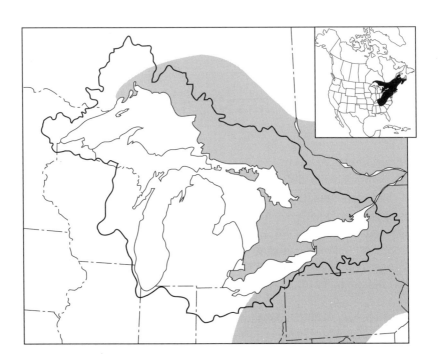

occupies grassy areas, bogs, swamps, and talus slopes, but its preferred habitat is deciduous or evergreen woods. The forest floor in smoky shrew habitat is shaded and moist and contains moss-covered logs or rocks and an abundance of plant litter overlying loosely packed soil. In favorable habitats, this species may amount to 20% of all the small mammals trapped. The smoky shrew, however, has a spotty distribution—abundant in some areas yet often absent in nearby, seemingly identical patches of woods. Also, it is scarce in areas where either the short-tailed shrew or the masked shrew is common.

The smoky shrew builds a spherical nest, about 23 cm (9 in) in diameter, from dry leaves and grass interspersed with a bit of fur. Nests are located in decaying logs, in tunnels, or under forest debris about 10–25 cm (4–10 in) below the surface. When active, this species occasionally runs about aboveground, but it most frequently travels through closed galleries within the leaf litter or inside mole tunnels. A smoky shrew forages throughout the day and night, at all times of the year, while continuously twitching its nose and uttering a faint twitter.

It is similar to other shrews in preying on a variety of invertebrates, including adult insects, caterpillars, grubs, centipedes, spiders, and earthworms. Some vertebrates, such as salamanders, occasionally fall prey to the smoky shrew, and an additional 15% or so of the diet is vegetable matter. In captivity, the smoky shrew eats up to half its body weight in food each day. Captive shrews also drink very little, which suggests that most of their daily water requirement is satisfied by metabolic water or water contained in their food.

Side (flank) glands of males become active at the onset of sexual maturity in spring and produce a strong musky odor that females presumably find attractive. Mating begins in late March and continues until late September. After a gestation of about 20 days, females give birth to an average litter of five young that are naked, blind, and helpless. How long the youngsters remain with their mother is not known for certain, but it must be short; some females mate soon after giving birth, and presumably a mother weans her first litter before the second is born. An adult female raises 1–3 litters each year, if she lives long enough.

Predators include hawks, owls, weasels, bobcats, red foxes, and gray foxes. Other shrews, particularly the short-tailed shrew, also prey on this species. Maximum lifespan is probably 14–17 months.

Suggested Reference: Owen 1984.

Pygmy Shrew *Sorex hoyi*

Measurements: Total length: 80–91 mm (3.1–3.6 in); tail length: 27–33 mm (1.1–1.3 in); hindfoot length: 8.0–10.5 mm (0.3–0.4 in); weight 2–4 g (0.07–0.14 oz).

Description: In summer, the dorsal fur of a pygmy shrew is grayish brown, and the underside is gray mixed with black; in winter, the back is less brown and more gray. The sparsely haired tail appears dark brown on top and gradually becomes lighter along the sides and bottom. Pygmy shrews, on average, are the smallest mammals in North America; however, large pygmy shrews are quite similar to small masked shrews. Compared to a masked shrew, a pygmy shrew tends to have a shorter tail and hindfoot and a less-protruding snout, but correct identification often requires an examination of the teeth. When viewed from the side, a pygmy shrew has only three obvious unicuspids; two others are present but not readily visible. In contrast, the first four unicuspids of a masked shrew are easily seen from the side (see fig. 30 in Key to Skulls).

Natural History: The pygmy shrew is widespread across the northern part of the continent and ranges southward through the Appalachian Mountains, mostly at higher elevations. This shrew lives in a bewildering array of sites including deciduous woods, coniferous forests, regenerating clear-cuts, grassy fields, swamps, bogs, and floodplains. Although the type of vegetative cover varies, most pygmy shrews live in moist boreal habitats with extensive ground cover. It is seldom abundant, and one estimate from northern Michigan indicates a density of only 0.5/ha (0.2/ acre) or less than one-tenth the typical density of the masked shrew. This apparent scarcity partly reflects the pygmy shrew's avoidance of standard live traps and snap traps; pitfall traps (large tin cans open at the top and buried flush with the ground) are much more effective at capturing this tiny mammal.

Like all shrews, the pygmy shrew is active year-round. It forages through leaf litter and the upper soil following tunnels that it builds itself or using those previously excavated by voles, moles, or even invertebrates, such as large beetles. In winter, it also travels along the surface of snow and in subnivean passageways. Captive animals have prolonged

Pygmy shrew. (Photo by Robert E. Wrigley, Manitoba Museum of Man and Nature.)

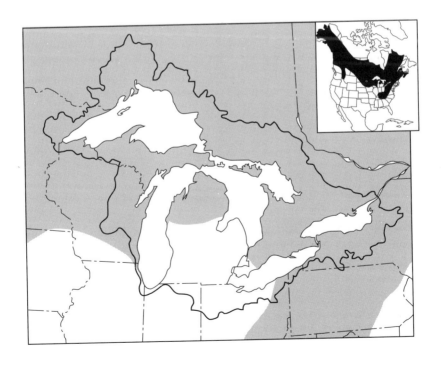

periods of rest punctuated by bursts of activity that average only three minutes long. During these active times, the shrew scurries about with its tail held stiffly behind and curved slightly upward.

In nature, a pygmy shrew preys heavily on spiders, caterpillars and grubs, adult beetles, ants, flies, and other insects. Presumably this mammal also attacks snails, slugs, and earthworms, as do most shrews. When invertebrate food is scarce during winter, this insectivore adds seeds to its diet. In captivity, it eagerly consumes dead white-footed mice and southern red-backed voles, and perhaps it scavenges such carcasses in the wild as well.

Biologists know little concerning reproductive patterns and longevity of the pygmy shrew in the Great Lakes region. Flank glands are well developed in males from June to August, suggesting that this period corresponds to the breeding season. If so, the mating season for the pygmy shrew is at least three months shorter than in other species, such as the masked and smoky shrews. Each female apparently produces only a single annual litter, and this small reproductive output also partly explains the low abundance of pygmy shrews. The gestation period probably is short, most likely 18 days, as in the masked shrew. Litter size varies from 3 to 8. Although young of the year approach adult size by winter, they do not reproduce until the following summer.

Average longevity is difficult to determine because field mammalogists recapture very few pygmy shrews. The oldest known pygmy shrew from our area apparently is an individual from the Upper Peninsula of Michigan that lived for 11 months in the wild. Reported predators include snakes, hawks, and house cats.

Suggested References: Long 1974; Ryan 1986.

Southeastern Shrew *Sorex longirostris*

Measurements: Total length: 77–92 mm (3.0–3.6 in); tail length: 26–32 mm (1.0–1.3 in); hindfoot length: 9–11 mm (0.4 in); weight: 3–6 g (0.1–0.2 oz).

Description: The southeastern shrew is a small species with a brown to reddish brown coat that becomes grayish on the belly. The tail is not distinctly bicolored. The range of the southeastern shrew overlaps that

Southeastern shrew. (Photo by Thomas W. French.)

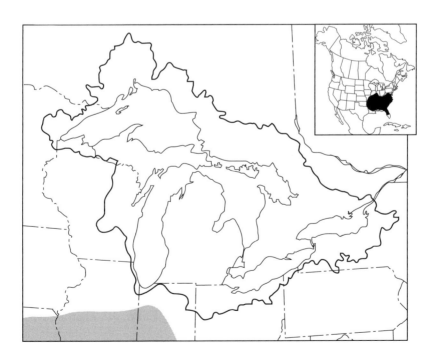

of the masked shrew, a species that is quite similar in appearance. Although the southeastern shrew usually has a smaller foot, shorter tail, and redder pelage, positive identification requires looking at the teeth. As in the masked shrew, the southeastern shrew has at least four obvious unicuspids when viewed from the side. However, the third unicuspid of a southeastern shrew usually is smaller than the fourth, whereas the third unicuspid is equal to or larger than the fourth in a masked shrew (see fig. 30 in the Key to Skulls).

Natural History: The southeastern shrew is primarily a southern species whose range only approaches the Great Lakes basin, near Lake Michigan, in northern Indiana and Illinois. Here, it is most abundant in moist to wet habitats adjacent to swamps, bogs, or rivers, although it also ventures into old fields, weedy fencerows, and even upland woods. Wherever it lives, the ground is well covered by dense grass, rushes, briars, or thick carpets of decaying leaves. The species is locally abundant and may attain population densities of 30–44/ha (12–18/acre).

The southeastern shrew is active both day and night, as are most shrews, and it rests in a nest of leaves hidden within a rotting log. True spiders and caterpillars make up 40–55% of the diet, but harvestmen, gastropods (slugs and snails), and centipedes each contribute another 10% or so. This insectivore preys on crickets, beetles, roaches, bugs, and earthworms to a lesser extent. The southeastern shrew generally supplements its meaty diet with small amounts of vegetation.

Average litter size is four, but ranges from one to six. Mating occurs between March and October, and young are born about three weeks later. In the southern part of its range, at least some females breed during the summer of their birth. House cats and owls, especially barred owls and barn owls, are the most common predators; the Virginia opossum, domestic dog, and various snakes also attack the southeastern shrew.

Suggested References: French 1980a, 1980b, and 1984.

Water Shrew   *Sorex palustris*

Measurements: Total length: 138–164 mm (5.4–6.5 in); tail length: 63–74 mm (2.5–2.9 in); hindfoot length: 19–20 mm (0.7–0.8 in); weight: 10–18 g (0.4–0.6 oz).

Water shrew. (Photo by Robert E. Wrigley, Manitoba Museum of Man and Nature.)

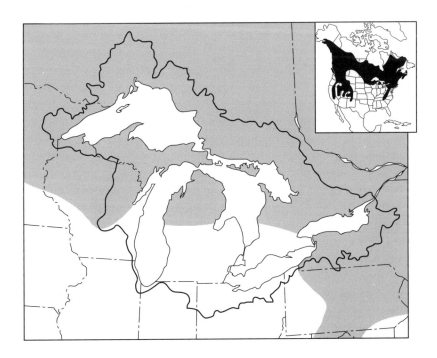

Description: This large shrew has dark gray to black fur and a moderately furred tail. However, the most distinctive characteristics of the water shrew are modifications of the feet associated with its semiaquatic lifestyle. The feet are large and have a stiff fringe of whitish hairs along the outer margin and between the toes; in addition, the third and fourth digits of the hind feet are partly webbed. Both modifications make the feet more efficient as swimming paddles, compared to those of more terrestrial shrews.

Natural History: The water shrew lives throughout much of the northern Great Lakes basin. As one might expect from its name, this species never strays far from open water. It occasionally resides near a sluggish stream, bog, or seasonal pond, but optimal habitat is a small, forest-lined stream, with fast-flowing water, and plenty of cover provided by undercut banks, jumbled rocks, downed trees, and other debris. Population density is generally low, and home range is a modest 0.2–0.3 ha (0.5–0.8 acre).

This species is active throughout the year and even swims underneath the ice of frozen creeks. The water shrew forages both night and day, and it usually alternates 30-minute activity bouts with periods of sleep lasting close to 60 minutes. When resting, this solitary animal occupies a leafy nest hidden inside an underground burrow or rotting log. Nests have an average diameter of about 8 cm (3 in), although some are as large as 30 cm (12 in).

A water shrew forages not only on land but also along the bottom of a stream or lake. The shrew's dense fur traps a large amount of air and allows this mammal to stay warm even when submerged in cool northern waters. The water shrew is capable of diving for up to 48 seconds but must work at staying submerged; the trapped air makes the animal extremely buoyant, and the shrew immediately bobs to the surface once it stops swimming. Trapped air bubbles also transform the appearance of this dark-colored mammal so that it appears as a fast-moving "silver bullet" when under water. It swims using a walking motion involving all four feet, and the specialized appendages also allow this intriguing mammal to run across pools of water, up to 1.5 m (5 ft) in diameter, without sinking.

The water shrew preys heavily upon aquatic larvae (nymphs) of certain insects, such as mayflies, stoneflies, caddisflies, and some of the true flies, and it also consumes leeches and snails. The shrew searches for such delicacies by continually probing the bottom mud and crevices between submerged rocks with its sensitive snout. Terrestrial invertebrate prey includes crickets, slugs, and earthworms. This shrew also feeds to a lesser extent on vertebrate flesh, such as adult and larval salamanders, fish eggs, and even minnow-sized fish. If a water shrew obtains surplus food during a particular foraging bout, it caches the excess in a secluded place and returns at a later time to finish the meal.

Compared to most shrews, the water shrew is an early breeder with mating activity starting in February, possibly even in January, and continuing until late summer. The gestation period is unknown but probably is similar to the 20–22 days observed in other large shrews. As many as 10 embryos develop, although a typical litter contains 5–7 offspring. Adult females produce 1–3 litters each season. A few young females, born early in the year, manage to reproduce late in their first summer; however, most defer breeding until the following year.

No data are available on longevity, but the water shrew probably does not live longer than 18 months in the wild. Fish, particularly trout, prey on water shrews, and other known predators include snakes, weasels, and hawks.

Suggested Reference: Beneski and Stinson 1987.

Least Shrew *Cryptotis parva*

Measurements: Total length: 64–86 mm (2.5–3.4 in); tail length: 12–18 mm (0.5–0.7 in); hindfoot length: 9–12 mm (0.4–0.5 in); weight: 4.0–6.5 g (0.1–0.2 oz).

Description: The least shrew is grayish brown on the top and sides and slightly paler underneath; winter-captured specimens are somewhat darker. The tail is lightly furred and very short, generally contributing less than 25% to the animal's total length. The short tail distinguishes this species from comparably sized, but more slender, masked shrews and pygmy shrews. The short-tailed shrew also has a short tail relative to its total length, but that species is much larger overall. In addition, the least shrew has fewer teeth (30) than any other shrew in the Great Lakes basin.

Natural History: The least shrew has a broad geographic distribution that includes much of the eastern United States, eastern Mexico, and

Least shrew

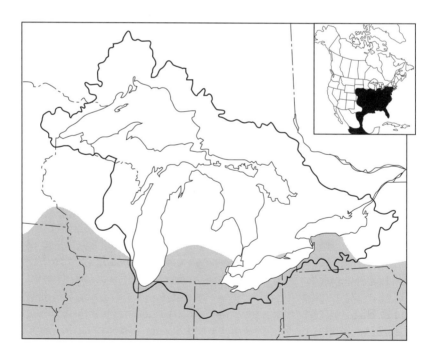

most of Central America. In Canada, however, the least shrew exists only at Long Point, Ontario; its presence here, and not near the Niagara or Detroit rivers, suggests that this small population resulted from individuals that drifted across Lake Erie from Pennsylvania or Ohio. Despite its wide distribution, the least shrew is rare throughout the Great Lakes basin; none, for example, has been captured in Michigan or Wisconsin since 1960.

Most shrews prefer moist or even wet habitats, but this species is an exception, favoring dry upland areas instead. The least shrew is most common in meadows, old fields, weedy areas, and along fencerows; a dense cover of grasses and/or herbaceous plants is always present. This shrew rarely visits woodlands or damp lowlands. Density is low, perhaps 2–5/ha (0.8–2.0/acre), and individuals restrict their activities to a home range of 1,700–2,800 m² (0.4–0.7 acre).

When foraging, it follows grassy runways built by the meadow vole or prairie vole, or it forms its own passageways through the thick vegetation; these are quite small, however, barely wider than a pencil, and easily overlooked even by experienced mammalogists. While inactive, the least shrew curls up in a nest woven from shredded grass, leaves, or corn husks. Nests often are tucked under a rock or log or beneath a piece of discarded lumber or metal sheeting; on other occasions, the least shrew places the nest, as much as 20 cm (8 in) below ground, at the end of a short burrow system. Each nest measures up to 14 cm (5.5 in) in diameter and typically has two entrances that lead into adjoining tunnels or runways. As a rule, shrews are highly aggressive toward one another, but this species is actually colonial at times; anywhere from 2–31 least shrews may share a single nest. This habit of communal nesting is more common in winter and probably is a means of conserving heat in a cold environment.

A least shrew usually subdues its prey with several swift bites to the head. The most important foods are adult insects, caterpillars, beetle larvae, and earthworms, but this insectivore also takes centipedes, slugs, and sowbugs. It prefers crickets to grasshoppers, eats more than its own weight in prey each day, and hoards excess food. This tiny mammal consumes small prey in their entirety, but it opens larger items, such as grasshoppers, and devours only the highly digestible, internal organs.

Mating occurs from March to November, and females give birth to a litter of 3–6 young, 21–23 days after copulation. Neonates are deaf, blind, and hairless and weigh only 0.3 g (0.01 oz); teeth are lacking, but tiny claws are apparent on the developing toes. Youngsters open their eyes and are fully furred by day 14; after 21 days, they weigh 4–5 g (0.1–0.2 oz), and weaning begins.

Although a captive individual lived for 21 months, least shrews in the wild probably are not so fortunate. Owls are the most consistent predators. The red fox, spotted skunk, house cat, and various snakes also eat an occasional least shrew.

Suggested References: Kivett and Mock 1980; Whitaker 1974.

Northern Short-tailed Shrew *Blarina brevicauda*

Measurements: Total length: 108–140 mm (4.3–5.5 in); tail length: 18–32 mm (0.7–1.3 in); hindfoot length: 13–18 mm (0.5–0.7 in); weight: 15–30 g (0.5–1.1 oz).

Description: The large size of the short-tailed shrew leads some beginners to confuse this species with a mouse or vole, but the dense coat, elongate snout, minute eyes, and pigmented teeth mark this mammal as a shrew. Dorsal fur is black, brownish black, or silvery gray, whereas ventral fur is lighter and generally more grayish. The tail is extremely short, amounting to less than 25% of the total length. Among shrews in the Great Lakes basin, only the least shrew has such a short tail, but that uncommon species weighs only one-quarter as much as a short-tailed shrew and is dramatically smaller in all linear measurements.

Natural History: The short-tailed shrew ranges throughout the Great Lakes region, and it is, perhaps, the most frequently captured of our shrews. It occurs in a variety of disturbed and undisturbed habitats— grasslands, old fields, fencerows, and marsh borders, as well as deciduous and coniferous forests. This large shrew prefers moist environments with extensive herbaceous cover or a thick layer of litter; it is, for example, less likely to live in a heavily grazed pasture or plowed field than in a mature deciduous woodlot. Densities generally range from 5 to 30/ha (2 to 12/acre) but exceed 200/ha (80/acre) under exceptional circumstances.

The short-tailed shrew rests in a hollow nest of leaves and grass that occasionally is lined with fur from a deceased meadow vole. These

Northern short-tailed shrew

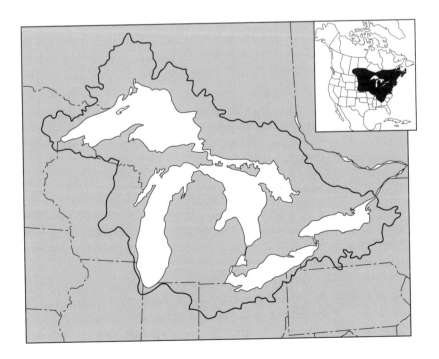

comfortable beds are located under downed trees or in subterranean cavities. Nests are quite large, up to 20 cm (8 in) in diameter, and generally have two or three exits that lead into various tunnels. Specific areas inside the burrow system act as "latrines" for the accumulation of fecal waste, whereas other chambers serve as storage sites for surplus food items.

The short-tailed shrew spends less time aboveground than other species, preferring instead to travel in shallow tunnels through the leaf litter. Activity generally occurs during frenzied five-minute bouts that, when added up, amount to only four hours each day. Foraging primarily takes place during the first two or three hours after sunset, although daytime activity often occurs, especially on cloudy days. When moving about, this solitary mammal relies on a fine-tuned sense of touch and the ability to echolocate; its eyes are rudimentary, create no images, and only sense light versus dark.

Although it eats subterranean fungi and seeds in small amounts, the short-tailed shrew is mostly carnivorous. It preys heavily upon insects and earthworms but also takes centipedes, millipedes, spiders, slugs, and snails. The short-tailed shrew includes vertebrates in its diet more often than other shrews and readily attacks smaller shrews, voles, mice, salamanders, and snakes.

Large size and an aggressive nature are sufficient to make the short-tailed shrew a formidable predator, but it is also one of the few venomous mammals in the world. Using a modified salivary gland, the shrew produces a potent poison that flows along a trough, formed by the lower incisors, and into the struggling prey. The toxin results in paralysis and may cause death in mouse-sized animals due to respiratory failure and circulatory malfunctions. Humans sometimes experience swelling and irritation after the bite of this shrew.

Mating occurs anytime from March to September, but most females give birth early or late in the season. Litters of 4–7 offspring appear after a gestation of 21–22 days. Young short-tailed shrews weigh less than a gram (0.04 oz) and are hairless, pink, and blind at birth. They live off mother's milk for up to 25 days, and a few become sexually mature after only 2–3 months.

Although one captive short-tailed shrew lived for 33 months, the average life span in the wild undoubtedly is much less. In one study, only about 6% of the animals marked and released one summer were still alive the following year. Cold stress may contribute to such high mortality, but predators also take a huge toll. Trout, sunfish, snakes, hawks, owls, foxes, weasels, skunks, dogs, and my neighbor's cat are just a few of the documented predators. Many mammalian carnivores seem loathe to eat a short-tailed shrew, even after killing it, perhaps because of the shrew's distinctive musky odor (and presumably taste).

Suggested References: George, Choate, and Genoways 1986; Martin 1981.

Moles
Family Talpidae

Moles occur only in the Northern Hemisphere, primarily in temperate areas of Eurasia and North America. This family was never as diverse as the shrews, and today it contains only 42 living species classified into 17 genera. One-sixth of all moles live in North America, and three species, each representing a different genus, inhabit the Great Lakes basin.

Among talpids, a few species are totally terrestrial (e.g., Asiatic shrew mole), and others have webbed feet and a semiaquatic life-style (e.g., Russian desman). However, the majority of species are exquisitely adapted for a subterranean existence. Massive claws at the end of each toe loosen hard-packed soil, and huge, paddlelike forefeet act as shovels. The animal moves the loosened dirt using a powerful sideways motion of the forelimbs, and, consequently its hands are turned permanently outward rather than down. The upper arm bone (humerus) is short, yet robust, and provides a large area for the attachment of muscles needed for digging. Football players and moles share the "no-neck" look, which is attributable to the massive development of shoulder muscles.

In addition, individual hairs in the velvetlike coat are short, equal in length, and lie in any direction, thus allowing a mole to move easily, forward or backward, inside narrow tunnels. Because moles inhabit the lightless underworld and rarely come aboveground, good vision is not a necessity; not surprisingly, their eyes are minute and, in some species, permanently covered by skin. Moles, however, compensate for their reduced vision with well-developed senses of touch and olfaction. External ears are lacking, and the ear opening is hidden beneath dense fur.

The skull of a mole is long, narrow, and flattened (fig. 5). The rear teeth display W-shaped ridges typical of other insect-eating mammals, including all shrews and bats in the Great Lakes region. However, a mole skull is at least 30% longer than that of any bat or shrew in our area. Also, mole skulls have a slender zygomatic arch that is missing in shrews, and moles do not have pigmented teeth like those found in shrews.

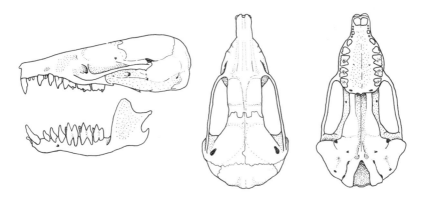

Fig. 5. Skull of an eastern mole, family Talpidae

Hairy-tailed Mole *Parascalops breweri*

Measurements: Total length: 150–170 mm (5.9–6.7 in); tail length: 23–36 mm (0.9–1.4 in); hindfoot length: 18–20 mm (0.7–0.8 in); weight: 40–64 g (1.4–2.3 oz).

Description: Broad forefeet, stout claws, minute eyes, and hidden ear openings easily identify this mammal as a mole. The dense, soft fur is dark brown to black across the back and over the triangular head. The conical snout is longer that that of any shrew and unfurred; lateral-facing, crescent-shaped nostrils are readily apparent. The tail is thick, constricted at its base, and appears scaly beneath the fur. Small body size and a tail covered with stiff black hairs distinguishes the aptly named hairy-tailed mole from other moles in the Great Lakes region.

Natural History: The hairy-tailed mole ranges along the mountains from Tennessee to Quebec and then westward to the Lake Superior shore. This species originally occupied deciduous woods and mixed forests of hardwoods and conifers, although today it also lives in old fields, grassy roadsides, and unused pastures. This mole prefers sandy loam soils with a fair amount of surface cover; it avoids soil that is high in clay and soil that is too dry or very wet. An average density is 3/ha (1.2/acre), but concentrations up to 30/ha (12/acre) occur.

Although the hairy-tailed mole ventures aboveground on warm nights, it spends most of its life underground. Just below the surface, the mole

Hairy-tailed mole

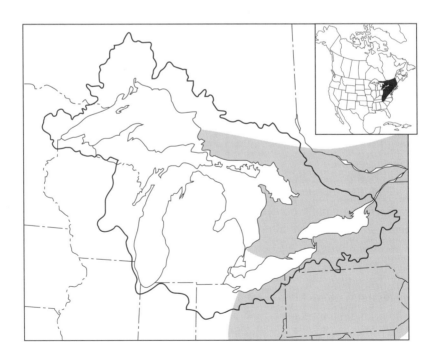

excavates a complex network of surface tunnels for foraging and commuting. Well-used passageways have smooth walls with solidly packed soil and measure 4 cm (1.6 in) wide and 3 cm (1.2 in) high. A hairy-tailed mole is not territorial, and other moles, as well as rodents and shrews, often use these underground highways. Some tunnels constructed by this species have been continually used for up to eight years—long after the original builder had passed on. Tunnel construction occasionally results in a ridge of upturned soil on the surface, but such ridges are difficult to see compared to those made by an eastern mole.

During winter, a hairy-tailed mole restricts its activity to deep tunnels, located 25–45 cm (10–18 in) below the frozen surface. Here, it builds a bulky warm nest from dried leaves and grass. Soil displaced while excavating a deep tunnel eventually appears as a molehill at the surface; those produced by the hairy-tailed mole are small and measure about 15 cm (6 in) across and 7.5 cm (3 in) high. Molehills appear most often in autumn as these mammals prepare deep tunnels for the coming winter.

A hairy-tailed mole is active day and night in an almost continual search for food. This mole preys mostly on earthworms and insects, yet centipedes, millipedes, snails, and slugs are also part of the diet. Beetles, both larvae and adults, are the most common insect prey. Although a hairy-tailed mole often ingests small amounts of plant roots, this species cannot survive on a strict vegetarian diet. A hairy-tailed mole eats from one to three times its own body weight in food each day.

Mating takes place in early spring and 3–6 young are born in a leafy nest, after an estimated gestation of 4–6 weeks. The wrinkled-looking newborn are toothless and essentially hairless, and the eyes are barely visible as black dots beneath the skin. Although the forefeet are already enlarged, the claws are blunt, soft, and short. Youngsters stay in the nest for about a month, until they are weaned, and then begin exploring the underworld on their own. This mole reaches sexual maturity when 10 months old, and each adult reproduces just once each year.

Some hairy-tailed moles live up to four years, but the average life span is probably much less. Reported predators include the red fox, Virginia opossum, domestic dog and cat, various snakes and owls, and even the bullfrog. The short-tailed shrew may attack nestling moles.

Suggested Reference: Hallett 1978.

Eastern Mole   *Scalopus aquaticus*

Measurements: Total length: 150–200 mm (5.9–7.9 in); tail length: 20–38 mm (0.8–1.5 in); hindfoot length: 22–27 mm (0.9–1.1 in); weight: 65–140 g (2.3–4.9 oz).

Description: A short, mostly naked tail and a plump body are the trademarks of this mole. Coat color varies from black to gray to brown, often with a silvery sheen. The large forefeet are broader than long, and the huge claws are almost as long as the fleshy portion of the foot. Toes are webbed, and the soles are surrounded by a fringe of stiff hairs. In contrast to the hairy-tailed mole, the opening of each crescent-shaped nostril faces upward.

Natural History: Although this species has the widest geographic distribution of any mole in North America, it occupies only the southern part of the Great Lakes basin. The webbed feet with a hairy fringe, reminiscent of a water shrew, led early biologists to believe that this mole lived in wet environments; even Linnaeus was fooled when he coined the name *aquaticus*. However, the eastern mole does not willingly enter open water and even avoids wet soil when burrowing. This large mole lives beneath forests, fields, pastures, and, to the dismay of many homeowners, lawns and gardens. It prefers slightly damp earth with a loamy or sandy texture and avoids rocky soils and those high in clay.

When excavating shallow foraging tunnels, an eastern mole loosens the dirt with powerful outward sweeps of the forefeet then turns on its side and pushes upward. This procedure raises and cracks the soil surface, leaving behind a telltale ridge. Under good conditions, an eastern mole produces tunnels at a rate of 5 m/hour (15 ft/hour). Tunnelling activity is particularly pronounced after the spring thaw or following a soaking rain during the otherwise dry days of midsummer.

The eastern mole uses deeper tunnels, about 25–75 cm (10–30 in) below the surface, during dry spells and especially during winter. When digging these tunnels, the mole either dumps excess soil into unused chambers or pushes it upward until it becomes a molehill—a circular mound of loose dirt, up to 30 cm (12 in) wide and 15 cm (6 in) high. Within the deep-tunnel system, the eastern mole carves out a large cavity, 20 cm (8 in) in diameter, and builds a nest of grass and leaves.

Eastern mole. (Photo by John and Gloria Tveten.)

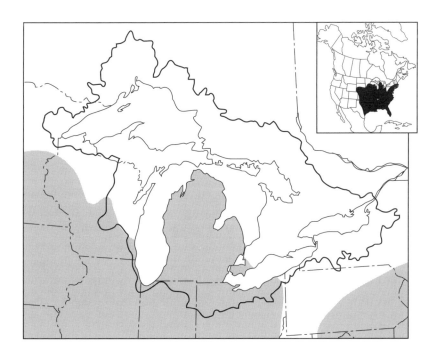

This species alternates foraging activity with periods of rest lasting 1–3 hours. About 25% of the diet is earthworms and another 40% comes from adult and immature insects, especially beetles and ants; additional prey include slugs, centipedes, and millipedes. In nature, this mole consistently eats roots and seeds, and captive individuals relish apples, potatoes, corn, wheat, oats, and tomatoes.

Eastern moles are solitary animals that come together for breeding in late March or early April. The gestation period is 40–45 days, and each female bears a single litter of 3–5 naked and toothless young in May. The young moles weigh about 5 g (0.2 oz) at birth, are furred by 10 days of age, and weaned after 4–5 weeks, although they are not ready to mate until the following spring.

An eastern mole is well protected in its subterranean habitat, and predation probably is not a major worry. Nevertheless, hawks and owls apparently swoop on this mole during those odd times when it does come to the surface, and dogs, cats, and foxes occasionally dig an eastern mole from its shallow foraging tunnel. Mammalian predators, however, often abandon the carcass before eating it; perhaps they find the musky odor of these moles repulsive. Longevity in the wild is not known, but a captive individual lived for almost two years.

Suggested Reference: Yates and Schmidly 1978.

Star-nosed Mole   *Condylura cristata*

Measurements: Total length: 175–205 mm (6.9–8.1 in); tail length: 65–85 mm (2.6–3.3 in); hindfoot length: 25–30 mm (1.0–1.2 in); weight: 35–75 g (1.2–2.6 oz).

Description: A hairless snout tipped with 22 fleshy tentacles makes the star-nosed mole one of the most distinctive of all mammals. The tentacles are small, just 1–3 mm (0.04–0.12 in) long, and evenly divided, with 11 on each side. The dense, dark fur appears almost black. Compared to the eastern or hairy-tailed mole, this species has a slimmer body, larger eyes, fur that is less velvetlike, and a tail that is twice as long. Large amounts of fat are stored in the tail during late winter, causing it to swell to three or four times its normal diameter.

Star-nosed mole. (Photo by Graham C. Hickman.)

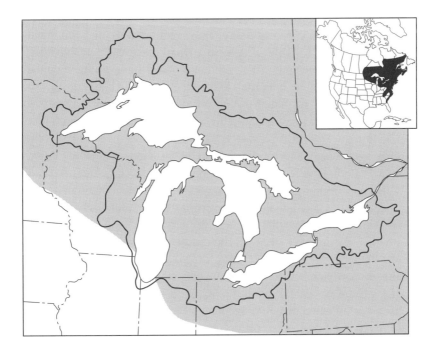

Natural History: The star-nosed mole ranges farther north than any other mole in North America. It lives throughout most of the Great Lakes region and is the only mole in much of the Lake Superior basin. Unlike other species, the star-nosed mole prefers wet, even saturated, soils, and it frequents the borders of swamps, lakes, streams, and isolated areas of poor drainage.

As do other moles, the star-nosed mole excavates shallow tunnels for foraging in summer and deep tunnels for protection in winter; however, this species has some very different habits compared to the eastern and hairy-tailed moles. For instance, the star-nosed mole often forages above-ground; its legs are proportionately longer than in the other species, allowing it to walk or run with greater ease. The star-nosed mole also continues to visit the surface throughout the winter and readily burrows through snow. In addition, this fascinating species spends much of its time in water, even swimming beneath ice in winter. Some of its subterranean passageways actually open under the surface of a lake or stream and provide safe access to aquatic foraging grounds.

Whenever this animal is active, the snout and tentacles are constantly probing and twitching. The function of the tentacles, or "star," is still somewhat of a mystery. Many biologists believe that it is a sensitive tactile organ, allowing the mole to feel its way through dark tunnels and murky water. One recent study, however, suggests that the star is an electroreceptor, capable of detecting small electrical fields given off by the mole's aquatic prey. If so, the star-nosed mole and the egg-laying platypus are the only mammals known to possess such an unusual ability.

Diet depends on habitat. A star-nosed mole without access to open water preys most heavily on earthworms and, to a lesser extent, on adult and immature terrestrial insects. Moles that live near large bodies of water take mostly aquatic worms and leeches and supplement these with the larvae of aquatic insects, such as caddisflies, stoneflies, and midges. It also eats small fish on occasion.

Both males and females reproduce for the first time when 10 months old. Copulation takes place in late winter or early spring, and most young are born during April or May. The estimated gestation period is 45 days. At birth, the eye is only a black dot, but the distinctive tentacles are already conspicuous. A newborn weighs 1.5 g (0.05 oz), gains an average of 1 g (0.04 oz) each day, and reaches 33 g (1.2 oz) by day 30, when it becomes independent. Average litter size is five.

Avian predators are mostly owls, ranging from the tiny saw-whet to the imposing great horned owl. Terrestrial enemies include dogs, cats, foxes, weasels, and skunks. The star-nosed mole is not safe in the water either, because the largemouth bass and bullfrog are reported predators. Maximum life span and average longevity are still unknown.

Suggested References: Gould, McShea, and Grand 1993; Petersen and Yates 1980.

Bats
Order Chiroptera

Bats have lived on the earth for over 50 million years, and today there are 926 living species organized into 177 genera and 17 families. One out of every five mammalian species is a bat. Bats live on all continents except Antarctica, and, in some areas such as New Zealand, they are the only indigenous mammals. They inhabit a diversity of environments and consume an amazing array of foods. Many bats eat insects, others catch fish or frogs, and a few subsist only on blood; many bats eat fruit, and some act like hummingbirds, lapping nectar from flowers. Economically, bats are extremely beneficial. They are important in pollinating plants, dispersing seeds, and controlling insect populations. They range in size from the tiny, 2-gram (0.07-ounce) bumblebee bat to the largest flying fox that weighs over 1.5 kg (3.3 lb) and has a wingspan of 1.7 m (5.5 ft).

Bats are the only mammals capable of true flight. This remarkable feat is possible because of extreme modification of the forelimb. Hand and finger bones, except the thumb, are very long and act as structural supports for the wing membrane. This membrane is a thin, soft, flexible piece of skin that runs along the body, from shoulder to ankle, and extends out between the fingers. The wing membrane is analogous to flight feathers in a bird, providing both lift against gravity and forward thrust for the flying bat. Many bats gain additional lift from a "tail membrane" (uropatagium) that passes between the legs and encloses the tail vertebrae.

Most bats rely on echolocation to find their way in the dark and to identify prey. When flying, a bat emits pulses of high-frequency sound (20–130 kHz) that generally are inaudible to humans. The bat listens for echoes bouncing off nearby objects and uses these echoes to determine the item's distance, size, shape, and surface texture. High-frequency sound does not travel far in the air; consequently, echolocation is useful only for detecting objects within a few meters of the animal. This inability to locate stationary obstacles or flying prey until the last minute contributes to the seemingly erratic flight of bats. Even though bats mostly depend on their ears, they do have eyes and see quite well. No bat is blind.

In recent years, newspaper and television reporters have given bats a sinister image, portraying these mammals as rabid and aggressive creatures. This is far from the truth. Although a few bats suffer and die from rabies, the vast majority, more than 99%, are free of this virus. The rare bat that contracts the disease usually becomes paralyzed rather than entering the highly agitated state typical of rabid dogs. In fact, most bites do not result from aggressive actions by the bat; most occur when a human picks up an injured or sick animal, and the bat simply tries to defend itself. Bats actually flee from humans that invade their roost and do not attack. A bat that zooms by your head on a warm summer's night is not being aggressive or trying to get into your hair. Why would it want to? It probably is chasing a mosquito intent on having you for dinner, and you should be grateful for the bat's help. The Chinese acknowledge the many beneficial aspects of bats and consider them symbols of good luck; maybe it is time for North Americans to adopt a similar attitude.

Plain-nosed Bats
Family Vespertilionidae

There are more than 300 species of vespertilionids making this the largest family of bats. The family has a cosmopolitan distribution, and almost all bats in the United States and Canada belong to it. Despite the huge number of species, all vespertilionids are quite small, with the largest weighing only about 50 g (2 oz). Eyes are generally tiny, but ears are well formed, spectacularly so in some species. In the spotted bat of western North America, for example, the ears are as long as the body. All vespertilionids also have a well-developed tragus—a fleshy projection at the base of the ear opening that aids in echolocation.

A W-shaped pattern on the cheek teeth of these bats (fig. 6) is indicative of the insect diet shared by most members of the family. Moles and shrews also have this dental pattern; however, a bat skull is much smaller than that of any mole in our area, and bat teeth lack the reddish brown pigmentation typical of our shrews. In addition, the skull of a vespertilionid has a unique horseshoe-shaped notch that clearly separates the upper teeth.

All vespertilionid bats in temperate areas share three interesting physiological processes—daily torpor, hibernation, and delayed fertilization. When entering torpor, a bat lowers its body temperature from the usual 35–38°C (95–100°F) down to the temperature of its surroundings. Torpor most frequently occurs after a cold or rainy night, conditions that greatly reduce the abundance of flying insects. Maintaining a high body temperature requires substantial energy, and torpor is simply the bat's way of conserving energy until food (insects) becomes abundant again. A torpid bat spontaneously raises its body temperature back to normal near sundown, and then flies off in search of a meal.

When autumn frosts spell the seasonal end of flying insects, bats enter hibernation. Hibernation is similar to daily torpor, but it differs in that the bat maintains a low body temperature for a prolonged period. Between October and April, the bat remains in hibernation, except for occasional arousals during which the animal urinates, drinks, or changes location. Arousals occur every 1–3 weeks and last for only a few hours each time. A bat survives the winter by slowly using up body fat that it stored the previous autumn; at the beginning of hibernation, fat may equal one-third of the bat's body mass.

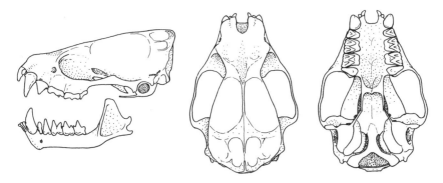

Fig. 6. Skull of a big brown bat, family Vespertilionidae

Temperate vespertilionids also practice "delayed fertilization." Although copulation typically occurs in September, a female does not become pregnant until the end of hibernation, seven months later. Sperm from an autumn mating actually is stored inside her uterus throughout winter. After leaving hibernation in the spring, the female finally ovulates, and a waiting sperm quickly fertilizes the egg. Spring is a time of low insect abundance, and the bat has depleted much of its fat and other nutrients over the long winter; delayed fertilization allows the animal to replenish its body without wasting precious time and energy on mating activities.

Small-footed Bat or Small-footed Myotis   *Myotis leibii*

Measurements: Total length: 73–82 mm (2.9–3.2 in); tail length: 30–35 mm (1.2–1.4 in); hindfoot length: 7–8 mm (0.3 in); ear height: 12–15 mm (0.5–0.6 in); forearm length: 30–34 mm (1.2–1.3 in); weight: 5–7 g (0.18–0.25 oz).

Description: This is a very small bat that weighs only as much as a 25-cent piece. It has glossy, yellowish brown or golden brown fur across the head and back and ventral hairs that are often much lighter. As in other bats, the naked ears and membranes are dark brown to black. The muzzle and skin around the eyes are black as well, and the species often is described as having a black "mask." Also, as one might expect, the small-footed bat does have a foot that is consistently shorter than that of

Small-footed bat. (Photo by Merlin D. Tuttle, Bat Conservation International.)

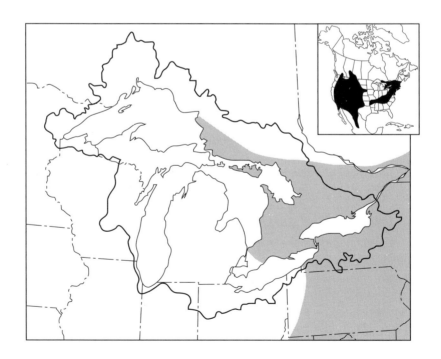

other species in the Great Lakes region. A small foot, short forearm, and black mask should separate it from other species.

Natural History: The small-footed bat has an irregular distribution across North America, and the eastern and western populations are probably distinct species. In the East, this bat ranges from Georgia and Alabama to Quebec. In the Great Lakes region, one is most likely to find this uncommon species in eastern Ontario, New York, and Pennsylvania.

The small-footed bat hibernates in mine tunnels or caves, often near the entrance in a dry, drafty location. It prefers a cool site compared to most other species and moves if air temperature rises above 4°C (39°F). More than 400 occasionally occupy the same cave, but the average number is less than 20. Unlike Indiana bats and little brown bats that cluster tightly together during hibernation, small-footed bats generally keep to themselves. Although most small-footed bats hang from the walls of their hibernaculum, some spend the winter underneath rubble in a mine or among fallen rocks in a cave. This hardy species is one of the last to enter hibernation and one of the first to leave in the spring. It subsists on stored fat throughout the long winter, and average body mass falls from 5.6 g (0.20 oz) in December to only 4.7 g (0.17 oz) in April, a loss of 16%.

We know little about this bat's habits in summer. Stomach analyses indicate that it preys on flies, beetles, bugs, leafhoppers, and flying ants. Apparently, males are solitary, and females form small maternity colonies consisting of fewer than 20 adults and their young. Most colonies find shelter within the crevices of a rocky cliff, although one was found in a crevicelike situation behind a sliding barn door. As in other *Myotis*, mating probably takes place at night, in a cave or mine during late summer, and females eventually become pregnant after leaving hibernation through the process of delayed fertilization. Birth occurs in June, and litter size is only one. This is a very small litter for most six-gram (0.2–0 ounce) mammals, but it is the most common litter size among bats.

As early as 1916, mammalogists began marking bats by placing a numbered metal band on the animal's forearm; these bands are similar to those used by ornithologists and allow one to recognize individual bats. Through recaptures of banded animals, biologists learn much about a species, including its migratory patterns and longevity. For example, banding studies indicate that the average annual survival rate for male small-footed bats is 76%, compared to only 42% for females. Maximum reported longevity for this species, determined by recapturing a banded individual, is 12 years. Band returns also suggest that small-footed bats travel only a short distance, perhaps less than 40 km (25 mi), between summer and winter quarters.

Suggested References: Barbour and Davis 1969; Fenton 1972; Hitchcock, Keen, and Kurta 1984.

Little Brown Bat or Little Brown Myotis *Myotis lucifugus*

Measurements: Total length: 80–95 mm (3.1–3.7 in); tail length: 31–45 mm (1.2–1.8 in); hindfoot length: 9–11 mm (0.4 in); ear height: 13–16 mm (0.5–0.6 in); forearm length: 36–40 mm (1.4–1.6 in); weight: 6–12 g (0.2–0.4 oz).

Description: The little brown bat has an evenly colored coat with a glossy sheen; it is olive brown to dark yellowish brown on the back and paler underneath. The black ears are shorter than those of the northern bat, and the blunt tragus is less than half the height of the pinna. Compared to the small-footed bat, the little brown bat lacks a black mask and has a larger foot. The little brown myotis has long hairs that extend beyond the tips of the hind claws, whereas the Indiana bat does not; in addition, the Indiana bat has a definite keel on its calcar, but the keel is absent or poorly developed in the little brown bat.

Natural History: The little brown bat occurs across most of the continent and ranges as far north as any bat in North America. In the winter, it hibernates in caverns and abandoned mines, where the air temperature remains between 2 and 5°C (36 and 40°F) and relative humidity is between 85 and 100%. Some mines contain more than 300,000 of these bats, although most have far fewer. Unlike small-footed or northern bats that are loners, the little brown bat forms loose clusters with others of its kind while hibernating. In April, this bat disperses to summer quarters located up to 350 km (210 mi) away.

After hibernation, males are solitary, roosting in tree hollows, under awnings, or behind shutters. Females, in contrast, form maternity colonies. Such colonies invariably are found in barns, houses, churches, etc., and consist of 150–300 or more adults. This species roosts in warmer

Little brown bat. (Photo by the author.)

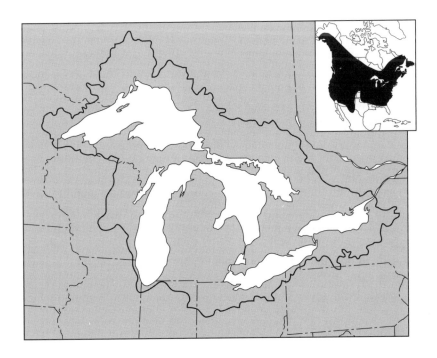

locations than the big brown bat, often choosing hot attics, where temperatures frequently exceed 38°C (100°F). A female is site loyal and consistently returns to the same maternity roost each year. The little brown bat is uncommon in urban areas and farmland typical of the southern Great Lakes basin, but it is the bat that most often uses buildings in the more forested north.

The little brown bat is active at any time of night, but most foraging takes place during the second and third hours after sunset. Even though it eats many types of insects, the little brown bat concentrates on those with an aquatic larval stage, such as mayflies and chironomid flies. Not surprisingly, it forages over streams and ponds, often flying just above the water's surface. Some individuals actually detect patches of prey by eavesdropping on the echolocation calls of other bats that are successfully feeding. A little brown bat consumes thousands of mosquito-sized prey each night, and during late lactation, an 8-gram (0.3-ounce) female eats 10 g (0.4 oz) of insects in a single night. This high food intake is necessary to meet the energetic demands of flight and milk production.

A female mates during late summer, stores sperm throughout winter, and ovulates in April after spring arousal. Average gestation is 60 days, but the actual length varies depending on the weather. On cold or rainy nights, there are few flying insects, foraging success is very low, and the bats respond by entering torpor. Although torpor lessens the demand for energy, the lower body temperature slows the rate of fetal growth; consequently, poor spring weather indirectly lengthens gestation by inducing torpor. At birth, the single pup weighs 2.3 g (0.1 oz); it is hairless and pink, with closed eyes and folded ears. A youngster grows rapidly and forages on its own after just 3–4 weeks.

Although a hawk or owl may pick off a flying bat, and a snake or raccoon might sneak into the roost for a quick meal, predation is not a big problem in this species. Exterminators, however, kill hundreds while evicting bats from houses; this is needless, because sealing the roost entrance, after the bats leave, is cheaper, safer, and more effective. Failure to store sufficient fat to last through the winter is a common mortality factor, particularly in the young; if fat supplies give out in February or early March, the bat dies of starvation because no insect food is available that early in the year. Little brown bats are the Methuselahs of the bat world, living up to 33 years in the wild. Males live longer than females.

Suggested References: Fenton and Barclay 1980; Kurta et al. 1989.

Northern Bat or Northern Myotis   *Myotis septentrionalis*

Measurements: Total length: 77–92 mm (3.0–3.6 in); tail length: 34–41 mm (1.3–1.6 in); hindfoot length: 8–10 mm (0.3–0.4 in); ear height: 16–19 mm (0.6–0.7 in); forearm length: 34–38 mm (1.3–1.5 in); weight: 5–8 g (0.2–0.3 oz).

Description: Fur of the northern bat is light to dark brown, and ears, wings, and tail membrane are naked and almost black. Large ears readily distinguish this mammal from other small-bodied bats. If a ruler is not available, gently lay the ears forward; in the northern bat, the ears extend at least 3 mm (0.1 in) beyond the tip of the nose, but less than this in other species. Furthermore, the long, slender, and sharply pointed tragus is a trademark of the northern bat; the height of the tragus amounts to more than 50% of the ear height.

Natural History: This species lives in most of eastern Canada and the United States. It hibernates in caves or abandoned mines where the temperature is usually between 2 and 7°C (36 and 45°F). Most often this animal hangs alone or in a cluster of just two or three bats. Also, it frequently tucks itself into drill holes or small cracks where it is easily overlooked. Hibernating populations are never large, and the most northern bats ever found in one hibernaculum is 304. A northern bat enters hibernation weighing about 8.5 g (0.3 oz), but body mass declines 30% by April as the bat steadily consumes its fat reserves.

After hibernation, a northern bat migrates to summer quarters located anywhere in the Great Lakes region; however, this bat is most abundant, in summer, in areas near the hibernacula. Males roost alone, while females gather in maternity colonies containing up to 60 adults. Some northern bats seek shelter in barns, behind house shutters, or under wooden shingles, but most shun man-made retreats. Using radiotracking techniques, one of my graduate students recently located 18 trees occupied by this species. His data indicate that northern bats have a preference for cavities within silver maples, although they also roost in hollow green ash and occasionally underneath the loose bark of dead trees.

The northern bat leaves the roost to forage every night soon after sunset. This species most often feeds within forests, below the canopy but above the shrub layer. It preys upon moths, beetles, bugs, caddisflies, stoneflies, and other insects. Although most bats in our area capture

Northern bat. (Photo by Merlin D. Tuttle, Bat Conservation International.)

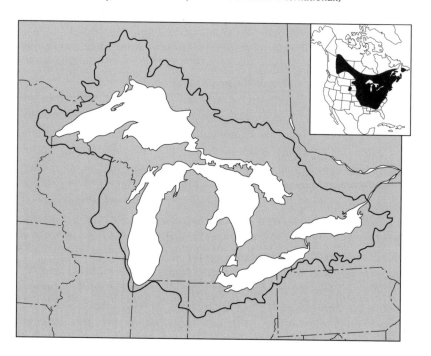

flying prey, some biologists believe that the northern bat is a "gleaner," plucking stationary insects from the surface of leaves or perhaps the ground. It detects prey using echolocation calls that have the highest frequency of any bat in the Great Lakes region, up to 110 kHz.

A female ovulates soon after leaving hibernation, and the egg is almost immediately fertilized by sperm stored in her uterus since September. After a development period of about 60 days, a single young is born at the maternity roost in June. The pup lives on its mother's milk for about 32 days before learning to fly and capture prey on its own. Maternity colonies break up during August and September when all bats migrate to hibernacula.

After leaving the summer roosts, a northern bat participates in "swarming," along with other species of *Myotis* and *Pipistrellus*. Although few bats use a cave or mine as a daytime roost in late summer, many make brief visits to the hibernaculum at night; so many bats attempt to enter that it seems as if one is surrounded by a swarm of large fluttering bees. Hundreds, and perhaps thousands, of bats flit about a large hibernaculum each night, and mating occurs inside during these nocturnal visits.

One of the most serious threats to the northern bat, as well as many other species, is the sealing of abandoned mines. Although such action prevents injury to humans, it also eliminates hibernation sites that are critical for the bats' survival. "Gating" the mine, i.e., closing it off with a network of steel bars, is a workable compromise, allowing bats to enter but not humans. There are no reports of predation on this species. Record longevity is 19 years in the wild.

Suggested References: Faure, Fullard, and Dawson 1993; Fitch and Shump 1979; van Zyll de Jong 1979.

Indiana Bat or Indiana Myotis *Myotis sodalis*

Measurements: Total length: 73–100 mm (2.9–3.9 in); tail length: 27–44 mm (1.1–1.7 in); hindfoot length: 8–10 mm (0.3–0.4 in); ear height: 12–15 mm (0.5–0.6 in); forearm length: 36–40 mm (1.4–1.6 in); weight: 6–11 g (0.2–0.4 oz).

Description: Basically, this is another small, drab, evenly colored species with blackish ears and membranes. It is most similar to the little brown

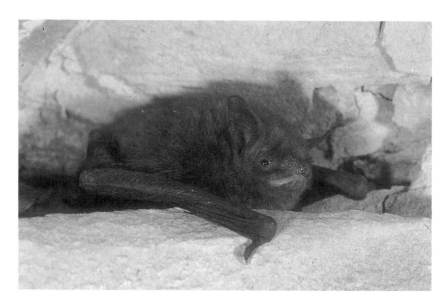

Indiana bat

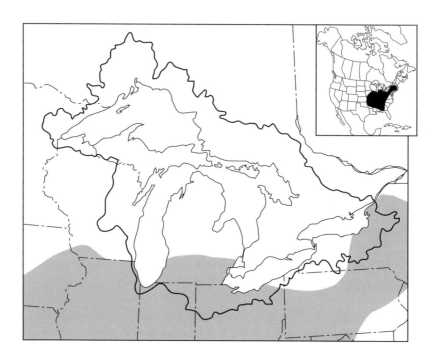

bat, and inexperienced biologists have trouble distinguishing between the two species. The Indiana bat, however, generally has dark brown to nearly black fur that appears dull rather than glossy. In addition, the Indiana bat has an obvious keel on its calcar and short hairs that do not extend past the tips of the hind claws.

Natural History: At one time, 85% of this species wintered in only seven different mines and caves, and two of these contained almost half of all the known Indiana bats. This lack of critical winter habitat is the reason that the United States Fish and Wildlife Service declared the Indiana bat an endangered species. Although this bat lives throughout much of the eastern United States, it does not occur in Canada, even though a small hibernating colony exists near Watertown, New York, at the eastern end of Lake Ontario, and summer populations are common in southern Michigan. This bat was recently discovered in Manistee County, Michigan, almost 200 km (120 mi) farther north than indicated on the map, but whether this record represents an accidental occurrence or is the result of the species expanding its range northward is unknown at this time.

This bat spends the winter in mine tunnels and natural caverns, most of which are far south of the Great Lakes drainage; major hibernacula, each containing 100,000 or more Indiana bats, exist in Indiana, Kentucky, and Missouri. These underground retreats have air temperatures of 4–8°C (39–46°F) and relative humidities of 75–95% throughout winter. Hibernating Indiana bats cluster very tightly together, more so than any other North American species, and they carpet the ceiling of their hibernacula in densities of 3,200 bats/m² (300 bats/ft²).

Most Indiana bats migrate into the Great Lakes basin beginning in late April. Once here, females form maternity colonies underneath the loose bark of dead trees; green ash, red oak, shagbark hickory, American elm, and cottonwood are common roost species. Most roost trees are in heavily canopied forests, but some are in pastures or open wetlands. Up to 100 bats occupy a single tree, although population size at any one roost fluctuates dramatically. For example, the population at one green ash tree in Michigan varied from 4 to 45 animals, as the bats shifted back and forth, among eight different trees, located within 160 m (525 ft) of each other. Females return to the same general area each spring and occasionally occupy the same tree that was used the previous summer.

Indiana bats begin leaving the roost to feed about 25 minutes after sundown. Preferred foraging habitat is a dense floodplain forest where a bat flies just above or below the canopy. A female wanders over a home range of 52 ha (128 acres) when pregnant and expands this to 94 ha (232 acres) after the birth of her pup. An Indiana bat eats flies, moths, and caddisflies that it captures in its mouth or plucks out of the air with a wing or tail membrane.

As in all our bats, ovulation and fertilization by stored sperm occurs in April, when females arouse from hibernation. Gestation is probably 60 days, similar to that of the little brown bat, and litter size is only one. Birth occurs in June, and most young learn to fly by late July. Starting in late August, Indiana bats begin migrating to southern swarming sites, but stragglers remain in northern areas until at least late September.

The biggest threat to this species is human destruction of hibernacula and summer roosting habitat. Predation is rare. In the wild, life span reaches 14 years.

Suggested References: Gardner, Garner, and Hofmann 1991; Kurta et al. 1993; Thomson 1982.

Red Bat *Lasiurus borealis*

Measurements: Total length: 93–117 mm (3.7–4.6 in); tail length: 40–50 mm (1.6–2.0 in); hindfoot length: 6–11 mm (0.2–0.4 in); ear height: 9–13 mm (0.4–0.5 in); forearm length: 36–46 mm (1.4–1.8 in); weight: 7–13 g (0.2–0.5 oz).

Description: This medium-sized bat is the most handsome mammal in the Great Lakes region and simply unmistakable. Overall the coat appears brick red to yellowish red, and the outer tip of many hairs is white, thus giving the red bat a frosted appearance; females have more of this "frosting." Unlike the naked tail membrane of most bats, that of a red bat is thickly furred on the dorsal surface. The snout is short, and the small, rounded ears are cloaked with thin reddish hairs. Most bats have only two nipples, but red and hoary bats have four.

Natural History: This fast-flying species lives throughout much of North and South America. In the Great Lakes basin, the red bat is abundant in rural areas and small towns but uncommon in heavily urbanized settings.

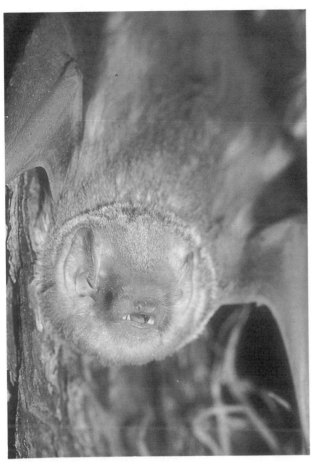

Red bat

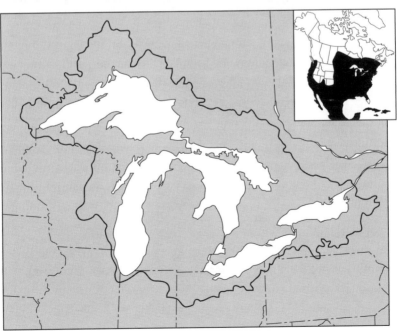

This bat is migratory, arriving in the Great Lakes region during late April and leaving by October, except for a few stragglers. During the winter, it probably hibernates in trees in areas to our south, such as the Ohio River valley.

The red bat is a solitary species that most often roosts in leafy trees, such as elms and maples, although it may hang in conifers as well. Roost sites are 1–6 m (3–20 ft) off the ground and hidden from view in all directions except below. The area beneath the roosting animal is clear of branches and allows the bat to take flight by simply dropping. A red bat uses its hairy tail membrane as an extra blanket to provide more insulation in cool weather.

Moths are common prey for red bats, but they also sample the occasional beetle, fly, leafhopper, or flying ant. A red bat forages for two hours each night, attacks an insect every 30 seconds, and succeeds in capturing its prey about 40% of the time. Total nightly consumption averages 6.2 g (0.2 oz), or almost half the bat's own body mass. A red bat forages along forest-field edges, over streams with some vegetative cover, and at lights.

It is fascinating to watch these bats capture moths attracted to a street light. The moth, which often can hear the echolocation calls of the bat, attempts to escape its aerial predator by quickly landing. The red bat, however, follows in a steep dive, often to within inches of the ground, before sharply pulling upward. A human walking beneath the light usually sees the diving bat but not the fleeing moth, and such a person often assumes that he or she is the object of the bat's aggression, thus giving rise to another false "bat-attacks-human" story.

Copulation takes place in August or September, and stored sperm finally fertilizes the eggs in March or April at the end of hibernation. Litter size is generally 2–3, although a red bat may give birth to as many as five young—the largest litter ever reported for a bat. Each newborn is hairless, has its eyes closed, and weighs about 1.5 g (0.05 oz). Most births occur in June, and young learn to fly in about five weeks. During the first few days after birth, a pup clings tenaciously to the fur of the mother and constantly seeks shelter beneath her wing; later, as the pup becomes stronger and more independent, it hangs next to her.

Hawks and owls are occasional predators, but blue jays also commonly harass and kill red bats. A Virginia opossum eats one now and then. In addition, some of these fast-flying mammals meet their end after darting into the path of a speeding automobile or after colliding with a radio tower or tall building. The latter fate seems more common during spring and autumn migration periods.

Suggested References: Hickey and Fenton 1990; Shump and Shump 1982a.

Hoary Bat *Lasiurus cinereus*

Measurements: Total length: 130–150 mm (5.1–5.9 in); tail length: 53–64 mm (2.1–2.5 in); hindfoot length: 10–14 mm (0.4–0.6 in); ear height: 17–19 mm (0.7 in); forearm length: 46–56 mm (1.8–2.2 in); weight: 25–35 g (0.9–1.2 oz).

Description: This is the largest bat in the Great Lakes region, with a wingspread up to 460 mm (18 in). Many hairs are tipped with white, making it appear as if the animal is coated with hoar frost, and the overall appearance is brown, yellowish brown, or tan, splashed with white. There is a yellowish brown throat patch that lacks the hoary tips. The tail membrane is totally furred on the dorsal surface, and the rounded ears are light brown inside and rimmed with black.

Natural History: The hoary bat also ranges across North America and down into South America. This strong flyer even lives on isolated oceanic islands, such as the Galapagos chain, and it is the only land mammal indigenous to Hawaii. It lives throughout the Great Lakes basin, where it is most common in rural areas and small towns but rare in cities. This large bat is only a seasonal resident, entering the Great Lakes region in April and leaving by early November. Where it spends the winter is still an unanswered question. Although some biologists speculate that hoary bats migrate to Mexico or Central America, it seems likely that they behave similarly to red bats and hibernate in the Ohio River valley or farther south.

Like the red bat, the hoary bat is a solitary species that roosts in a variety of trees, including maple, elm, cherry, and spruce. Probably any tree that provides dense shade, seclusion, and a clear space below the actual roost site is useable. A hoary bat roosts about 2–6 m (7–20 ft) above the ground.

Hoary bat

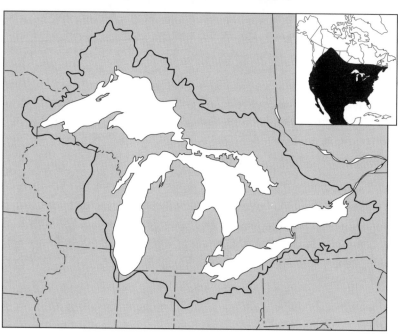

In most of our bats, activity peaks within the second and third hours after sunset, but the hoary bat appears to be a late flier, feeding mostly during the fifth hour after sundown. It forages over large canopied streams, along forest edges, and at lights; high speed and low maneuverability limit flight to fairly open and uncluttered situations. The hoary bat preys on large insects, detected from a distance, and individually pursued. Moths with body lengths of 14–29 mm (0.6–1.2 in) and beetles measuring 12–23 mm (0.5–0.9 in) are typical prey. When insects are scarce, a hoary bat defends a feeding territory 40–50 m (130–165 ft) in length; a territory owner pursues any intruder, makes audible vocalizations toward it, and occasionally bumps the stranger. The echolocation calls of a hoary bat have a lower frequency than those of most species in our area and are often audible to humans as a series of faint clicks.

Males reach breeding condition in late summer, and copulation presumably occurs at this time. Nevertheless, delayed fertilization results in most females not giving birth until June, after an estimated gestation of only 60–90 days. Normal litter size is two, even though this species has four mammary glands. Each newborn has a sparse coating of silvery hair and weighs only 4.5 g (0.2 oz). Ears become erect after 3 days, eyes open by day 12, and the pup learns to fly by 33 days of age. To meet the energy demands of her growing offspring, nursing mothers increase nightly foraging time by 73% between birth and weaning, and mothers with large young feed up to 6 hours each night.

Secluded roost sites, cryptic coloration, and fast flight make predation on this species uncommon, but occasionally one falls to a hawk, owl, or snake. Accidents happen to bats as well as to humans. For example, biologists occasionally find a dead hoary bat hanging on a barbed wire fence; presumably the flying bat strays too close to a barb, becomes snagged through a wing or tail membrane, and cannot escape. Some hoary bats live more than two years, but longevity information on solitary species is hard to come by.

Suggested References: Barclay 1989; Shump and Shump 1982b.

Silver-haired Bat *Lasionycteris noctivagans*

Measurements: Total length: 90–115 mm (3.5–4.5 in); tail length: 35–50 mm (1.4–2.0 in); hindfoot length: 6–12 mm (0.2–0.5 in); ear height: 15–17 mm (0.6–0.7 in); forearm length: 37–42 mm (1.5–1.6 in); weight: 8–12 g (0.3–0.4 oz).

Description: This medium-sized bat appears dark brown to black with a variable amount of white frosting. It could only be confused with the hoary bat, but the silver-haired bat is much darker and smaller and has a proportionately longer snout. Also, the dorsal surface of a silver-haired bat's tail membrane is only partly furred; the outermost 50–75% is naked.

Natural History: The silver-haired bat lives in most temperate areas of North America. It is a migratory species, entering the Great Lakes region in April and generally departing by November. However, a few silver-haired bats attempt to hibernate as far north as extreme southern Michigan.

Although it occasionally strays into buildings, this bat most often seeks shelter in tree hollows, underneath loose bark, or in narrow folds of heavily furrowed bark. It roosts about 1–5 m (3–16 ft) above the ground. The silver-haired bat seems particularly fond of willow trees but uses maple or ash trees as well. Most silver-haired bats lead a solitary life, although a dozen or so females occasionally form a colony.

The silver-haired bat is an early flier that forages over woodland ponds and streams or in small forest clearings. It is highly maneuverable, and its flight is so slow that one might mistake it for a huge moth. Flies, beetles, and moths are common prey items, but the silver-haired bat feeds opportunistically on any patch of insects that it comes across. The species avoids foraging in areas where the larger big brown bat is abundant.

Autumn mating, delayed fertilization, and a June birth is typical for bats in the Great Lakes region, and the silver-haired bat is no exception. Births occur after a gestation of 50–60 days, and litter size is generally two. The birth process is typical of vespertilionid bats. At parturition, a female abandons her usual head-down posture, roosts with the head facing upward, and curves her tail membrane forward to form a basket that catches the young as it leaves the birth canal. Breech birth is the norm, and a youngster actually helps by pushing against the mother with its hindfeet in an attempt to free the head and shoulders. Birth of each

Silver-haired bat

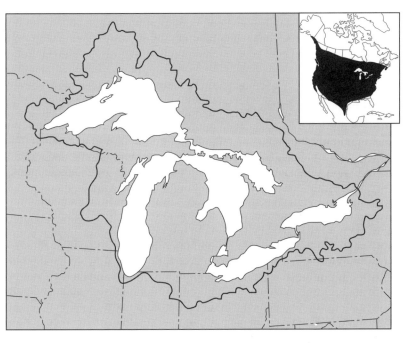

twin takes only about 20 minutes with a 20-minute rest period between the first and second birth.

A silver-haired bat is born without fur, with eyes closed, ears folded over, and most of its 22 deciduous ("baby") teeth already in place. Body skin is pinkish, and the wings are mottled with tan and black. Individual newborns weigh about 2 g (0.1 oz), and the combined weight of the litter equals 36% of the mother's body mass. Starting soon after birth, a pup emits high-pitched chirps whenever separated from its parent. The young are able to forage for themselves after about 36 days.

The only reported predators are the great horned owl and striped skunk. At twilight, surprised fishermen occasionally snag this bat in mid-air as they cast for trout; the bat apparently mistakes the baited hook for a flying insect. Studies based on tooth-wear suggest that the silver-haired bat lives up to 12 years and that it probably has a shorter life span than more colonial species, such as the big and little brown bats.

Suggested References: Barclay, Faure, and Farr 1988; Kunz 1982; Kurta and Stewart 1990.

Evening Bat *Nycticeius humeralis*

Measurements: Total length: 86–103 mm (3.4–4.1 in); tail length: 33–40 mm (1.3–1.6 in); hindfoot length: 8–10 mm (0.3–0.4 in); ear height: 11–14 mm (0.4–0.6 in); forearm length: 34–38 mm (1.3–1.5 in); weight: 6–12 g (0.2–0.4 oz).

Description: The evening bat is a plain-looking mammal with dark brown fur and blackish ears, snout, and membranes; wing and tail membranes are hairless. This species looks like a big brown bat but is much smaller in all dimensions. Compared to a little brown or Indiana bat, the tragus of an evening bat is very rounded and short, usually less than 30% of the ear height. In addition, an evening bat has only one pair of upper incisors, but a big brown bat and all *Myotis* have two pair.

Natural History: This is really a southern species that reaches the northern limit of its range in the Great Lakes basin. There is only one record of the evening bat from Ontario (Point Pelee) and three from southern Michigan, and the species apparently is rare to nonexistent in adjacent

Evening bat. (Photo by Gerald S. Wilkinson.)

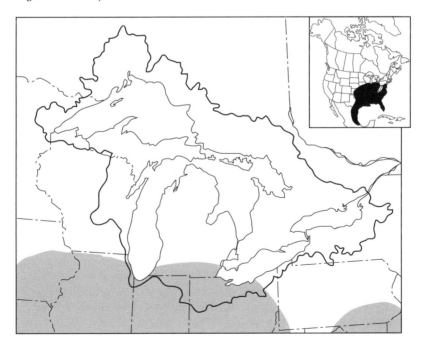

parts of northern Indiana and Ohio as well. The few specimens recorded from the Great Lakes area probably represent stray animals and not a resident population.

Evening bats live in rural areas or small towns, where females form summer maternity colonies in buildings or tree hollows. In southern Indiana, most colonies harbor fewer than 100 adults. Pregnant females begin arriving in early May, and a colony reaches peak size by the end of the month. Juveniles begin leaving the natal roost by late August, and everyone departs by mid-September. All evening bats apparently spend the winter south of a line stretching from South Carolina to Arkansas. In late autumn, an evening bat stores large amounts of fat, but whether it uses the fat to fuel its long migration or an extended period of hibernation is unknown. Adult males remain in the southern United States throughout the year and do not make the northward migration in the spring.

The prey of evening bats occur in rich patches that exist in different locations on different days. Individual bats that have not located a good feeding area often follow other, more successful bats to their foraging grounds. The evening bat feeds on beetles, moths, flies, and leafhoppers that the bat catches in midair during a slow, steady flight. A common prey item is the adult form of a chrysomelid beetle better known to farmers in its larval stage as the corn rootworm, a serious agricultural pest. A colony of 100 evening bats consumes over 1.25 million insects in a single season.

All females in a colony give birth within a six-day period in June. As in many bats, the mother snips the umbilical cord with her teeth soon after birth and ultimately eats the cord and placenta. A mother leaves her young in the roost during nightly foraging and uses auditory and olfactory cues to recognize her offspring when she returns. If a young pup falls from its precarious perch high in a barn, the mother descends to the ground and retrieves her flightless offspring.

Late in the lactation period, a female often allows an unrelated pup to nurse. Such unusual behavior may be a way of dumping excess milk that was produced by the mother but not consumed by her own offspring. Milk-dumping presumably decreases the weight that a female must carry on her next foraging trip, and emptying the mammary gland apparently promotes high milk yields. At peak lactation, this bat secretes half its body mass in milk each day.

Most females bear twins, but some produce triplets and successfully raise these large litters to maturity. Each neonate weighs about 2 g (0.07

oz), and the entire litter represents 50% of the mother's postpartum body mass—the largest litter in relation to maternal size of all bats and one of the largest for any mammal. Each pup is hairless and pink at birth; its eyes open within 30 hours of parturition. A young evening bat makes test flights within the roost at 20 days of age and starts foraging a few days later.

House cats often pounce on evening bats during nightly emergence, but there are no other documented predators; presumably snakes, raccoons, owls, and hawks occasionally prey on this bat. Humans sometimes kill evening bats while trying to evict them from a building roost. Average life span is two years, but a few live at least five years.

Suggested References: Watkins 1972; Whitaker and Clem 1992; Wilkinson 1992a and 1992b.

Eastern Pipistrelle *Pipistrellus subflavus*

Measurements: Total length: 70–80 mm (2.8–3.1 in); tail length: 28–45 mm (1.1–1.8 in); hindfoot length: 7–10 mm (0.3–0.4 in); ear height: 13–15 mm (0.5–0.6 in); forearm length: 32–36 mm (1.3–1.4 in); weight: 4–7 g (0.1–0.2 oz).

Description: Coat color of this tiny species varies from golden brown to reddish brown. Dorsal guard hairs have a distinct tricolored appearance—dark at the base, yellowish in the middle, and dark again at the tip. Wing membranes are naked, but the anterior third of the tail membrane is lightly furred. Light-brown hairs occur on the top of the feet and toes and sharply contrast with the underlying black skin. The forearm is reddish, and it readily stands out against the black wing membrane.

Natural History: This bat barely enters Canada, lives in most of the eastern United States, and ranges far into Mexico along the Gulf Coast. It generally is absent from the northern and south-central parts of the Great Lakes basin. Apparently the Canadian shore of Lake Superior is too cold, and much of the southern Great Lakes area lacks suitable hibernacula. On rare occasions, pipistrelles have been found outside the area indicated on the map, near the southern shore of Lake Michigan and also far to the north in Manistee County, Michigan. However, these most

Eastern pipistrelle

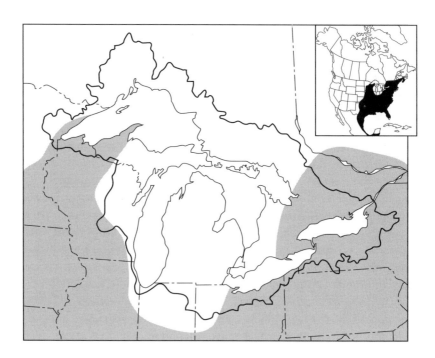

likely are accidental occurrences, perhaps windblown or disoriented individuals, that do not represent a resident population.

The pipistrelle hibernates in a cave or mine for 6–9 months each year. Anywhere from one to several hundred spend the winter in the same hibernaculum, usually hanging from the walls, not the ceiling, and generally in the warmest area available. Within the hibernaculum, an eastern pipistrelle roosts singly and does not huddle in the large clusters characteristic of little brown and Indiana bats. The pipistrelle is the first species to enter hibernation in autumn and the last to leave in spring. Body mass decreases 29–39% between October and April as the bat metabolizes its stored fat.

In summer, males roost alone, but females band together in small groups containing fewer than 20 adults. Eastern pipistrelles in southern states use caves as maternity roosts, but caves and mines in the Great Lakes area are simply too cold in the summer. Our pipistrelles consistently seek warmer sites, such as barns or possibly hollow trees.

This bat forages over streams and ponds and at the forest-field edge but avoids dense unbroken forests. It leaves the roost earlier than most bats, so early that it often flies among foraging chimney swifts. Despite a slow, erratic, flight pattern, the pipistrelle is an efficient predator, catching an insect every two seconds, and in just 30 minutes of foraging, the bat increases its body mass by 25%. The diet consists of small insects, 4–10 mm (0.15–0.20 in) long, and includes beetles, flies, leafhoppers, and the occasional moth or flying ant. This mammal flies continually while feeding and usually scoops unlucky prey out of the air with a wing or tail membrane.

A female is inseminated prior to hibernation, during autumn swarming, and stores the sperm inside her uterus until spring, when ovulation occurs. In June or early July, after a gestation of at least 44 days, each female gives birth to twins weighing 1.6 g (0.06 oz) each. Total litter mass represents 48% of the mother's mass after birth. Each neonate has the typical altricial look of vespertilionid bats—closed eyes, collapsed ears, and naked skin. As in all our bats, the thin deciduous teeth curve down and back and look very much like snake teeth; deciduous teeth are present at birth and presumably help maintain a firm grasp on the nipple. A young pipistrelle learns to fly in three weeks and is totally weaned by four weeks. Colonies disband by late summer, and the bats migrate a short distance, less than 50 km (30 mi), to a hibernaculum.

Eastern pipistrelles have few known predators. The only documented cases of predation are by a prairie vole and a leopard frog, but certainly these represent highly unusual, accidental encounters. Males live longer than females, and maximum reported longevity is 15 years in the wild.

Suggested Reference: Fujita and Kunz 1984.

Big Brown Bat *Eptesicus fuscus*

Measurements: Total length: 110–130 mm (4.3–5.1 in); tail length: 38–50 mm (1.5–2.0 in); hindfoot length: 10–14 mm (0.4–0.6 in); ear height: 16–20 mm (0.6–0.8 in); forearm length: 41–50 mm (1.6–2.0 in); weight: 15–24 g (0.5–0.8 oz).

Description: The big brown bat appears brown to reddish brown and is evenly colored across the dorsal surface. Ventral fur is distinctly lighter. The wing and tail membranes, ears, and muzzle are naked and black. In our area, this species is larger than any other bat, except the hoary bat, which has tan fur tipped with white.

Natural History: This species lives throughout temperate North America and follows the western mountains down through Central America into Colombia. In the Great Lakes region, it inhabits rural areas, towns, and cities. It is most abundant in country dominated by farmland and least common in heavily forested regions.

Throughout spring and summer, males are solitary, but females cluster in maternity roosts that usually shelter 20–70 adults. Such colonies typically reside in a house or barn, and this species is the most common building-dwelling bat in the southern Great Lakes basin. In houses, the bats most likely roost inside walls or a boxed-in eave and rarely enter attics. Unlike the little brown bat, the big brown bat does not relish a hot roost and generally relocates when air temperature exceeds 32°C (90°F). Females typically return to the same maternity site each year.

Many people object to sharing their home with the big brown bat or other species that commonly inhabit buildings. The only permanent solution to this conflict is to block the small holes that bats use to enter the building. Bats do not gnaw like rodents, and they cannot reopen a sealed entrance. Of course, sealing should be done only at night, after the bats leave to forage, and it should not be done during June or July when

Big brown bat

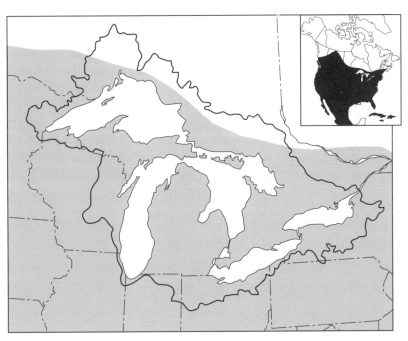

flightless young remain inside. Use of poisons is often illegal, dangerous, expensive, and ineffective, because other bats invariably reoccupy the roost the following year. Trying to evict bats with mothballs, sulfur candles, blaring radios, etc., does not work.

In good weather, colony members begin foraging about 20 minutes after sundown. Like most species, this bat does not feed in heavy rain or when air temperature is below 10°C (50°F). The big brown bat is a beetle specialist, consuming hundreds of these insects, including many crop pests, each night. This species often uses a "night roost"; after foraging, the bat hangs under a porch awning, behind a shutter, or in a barn to rest and digest its dinner. Before dawn, it returns to the day roost, leaving behind a small pile of droppings as the only evidence of its brief nocturnal visit.

Maternity colonies disband in August and September, and most bats migrate a short distance to winter quarters. Many big brown bats spend the winter in mines or caves, but this bat also hibernates inside the wall or attic of a heated building, where air temperature is cool but still above freezing. Dramatic changes in outside temperature often cause the bats to arouse and search for a more suitable site. A bat found in a building between November and March is almost certainly this species.

After a 60-day gestation, a female gives birth to one or two young in June. A pup is blind and naked and weighs about 3.3 g (0.1 oz). When separated from its parent, a young big brown bat makes a persistent chirping noise that humans readily hear and often mistake for a bird. As in other bat species, a mother carries her young in flight only if disturbed and leaves them in the roost when she forages. Bats, including the big brown bat, are unusual among small mammals in that the mother does not build a nest of any kind. A youngster forages on its own after 3–5 weeks.

Nocturnal habits and secluded roost sites protect these bats from most predators, although owls, snakes, raccoons, and house cats take a few. As in other hibernating bats, failure to accumulate enough fat for the long winter is a major mortality factor, particularly for inexperienced young. This species lives up to 19 years in the wild.

Suggested References: Kurta and Baker 1990; Whitaker and Gummer 1992.

Rabbits, Hares, and
Their Allies
Order Lagomorpha

The Lagomorpha is a small group of mammals that first appeared in the fossil record of Asia about 60 million years ago. Mammalogists currently recognize 81 living species, 13 genera, and just two families, the Leporidae and Ochotonidae. Members of the Leporidae typically have huge hind feet and long ears and include the familiar rabbits and hares. Humans have introduced this family into Australia, and it occurs naturally on all other habitable continents. In contrast, the Ochotonidae, or pikas, inhabit only part of Asia and mountainous regions of western North America. They are plump-looking animals, more similar in appearance to guinea pigs than rabbits; pikas have an inconspicuous tail, and their ears are not greatly elongated.

This order is superficially similar to rodents in that the incisors are well developed and separated from other teeth by a long gap or diastema. In both groups, the incisors continue to grow throughout life, keeping pace with constant wear resulting from the animal's gnawing habits. Lagomorphs differ, however, in having two pairs of upper incisors, unlike the single pair in rodents. One might easily overlook this second pair because it is quite small, peglike, and located directly behind the first pair, an arrangement unique among mammals (fig. 7).

These mammals have a number of other interesting characteristics. For example, the scrotum hangs in front of the penis and not behind it; only the marsupials share this anatomical trait. Lagomorphs also have extremely thin skin, and, unlike most mammals, hair covers the soles of

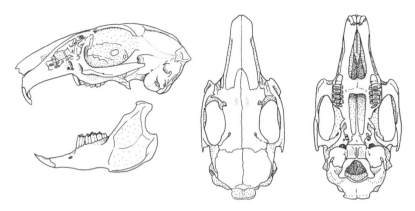

Fig. 7. Skull of an eastern cottontail, family Leporidae

the feet. In addition, many small openings, or fenestrations, occur in the skull, particularly in the Leporidae; note, for example, the area in front of the eye orbit shown in figure 7.

Lagomorphs also possess an intriguing digestive adaptation. These herbivorous mammals come equipped with an extensive cecum, an out-pocketing of the large intestine. Here, a multitude of symbiotic bacteria break down tough plant material that the animal could not digest in its stomach or small intestine. Although bacterial action releases nutrients into the cecum, neither the cecum nor large intestine can absorb these useful molecules. A lagomorph solves this problem by practicing co-prophagy. After the cecum packages the material into soft greenish pel-lets, the animal defecates; it then reingests the pellets, and the small intestine easily absorbs the nutrients during their second pass through the system. The animal only reingests green pellets, and it ignores hard brownish ones that contain undigestible material.

Rabbits and Hares
Family Leporidae

About 70% of lagomorphs are in this family. Their long ears, large hindfeet, and fluffy tail are known to all. Around the globe, humans have domesticated a few species and hunt many others for their soft fur and edible meat. Some rabbits and hares that escaped from captivity now flourish in areas far from their original range, and humans have intentionally transplanted others to provide additional game animals. Leporids currently live on all continents except Antarctica, and they occupy habitats ranging from deserts to moist tropical forests and from sea level to alpine meadows. Two species of rabbit and two hares occur naturally in or near the Great Lakes region, and another species of hare lives here only through human intervention.

What is the difference between a rabbit and hare? In terms of physical characteristics, the two are quite similar. They differ somewhat in size; for instance, hares are generally larger, weighing 1.3–7.0 kg (3–15 lb) compared to only 0.25–2.50 kg (0.5–5.5 lb) for rabbits. In addition, hares always have black-tipped ears, but rabbits often do not.

The most significant differences between rabbits and hares concern development and care of the young. Rabbits are relatively altricial; they are born naked or lightly furred, blind, and helpless. Hares, in contrast, are more precocial; neonates are well-furred, can see quite well, and run about soon after birth. In addition, rabbits always build a comfortable nest to shelter their offspring, but hares never do. Although biologists generally agree on these distinctions, the common names of some species lead to unfortunate confusion. As an example, the white-tailed jackrabbit of the Great Plains is actually a hare and not a rabbit at all.

Eastern Cottontail *Sylvilagus floridanus*

Measurements: Total length: 375–475 mm (15–19 in); tail length: 35–70 mm (1.4–2.8 in); hindfoot length: 80–110 mm (3.1–4.3 in); ear height: 50–65 mm (2.0–2.6 in); weight: 0.9–1.8 kg (2–4 lb).

Description: Hairs along the back are yellowish brown tipped with dark brown to black, whereas ventral hairs appear white with a dark gray base. The throat is tan. A large, rusty brown patch covers the back of the neck, and a white forehead blaze may be present. The powderpuff tail is dark above and white below. Although the Appalachian cottontail is very similar in color, it usually has a black spot between the ears and is slightly smaller, on average, than the eastern cottontail. Hares are heavier, have a considerably larger foot, and lack the rusty brown coloring on the neck.

Natural History: This cottontail barely enters Canada, but it is widespread across the United States east of the Rocky Mountains. The eastern cottontail is abundant where herbaceous vegetation abounds and potential shelter exists in the form of brushpiles, shrubby thickets, or weedy fencerows. It avoids extensive grasslands without suitable hiding places and deep forests with sparse groundcover. Consequently, the eastern cottontail is very common in the southern Great Lakes region where a mosaic of croplands, pastures, meadows, and woodlots exist. This rabbit, however, is scarce in northern Wisconsin and Michigan and absent from Ontario north of Lakes Huron and Superior, where large forested tracts dominate the landscape. Favorable habitats support up to 8 rabbits/ha (3/acre).

Home range is usually less than 2 ha (5 acres); it is larger in males than females and increases in size during the breeding season. These solitary animals forage at any time of day or night but perhaps more commonly just after sunrise and near sunset. A cottontail generally rests in a "form." This is nothing more than a shallow depression in the ground that is hidden beneath a pile of brush or a clump of long grass. During winter, some eastern cottontails occupy a burrow abandoned by a badger or woodchuck.

The eastern cottontail is a strict vegetarian that feeds from an extensive menu. In summer, grasses form the bulk of its diet, but it also eats clover, plantain, dandelion, goldenrod, and wild carrot. To the dismay of many gardeners, this rabbit relishes beans, peas, lettuce, and other delights

Eastern cottontail

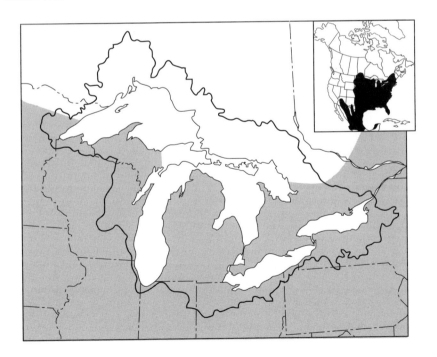

intended for the human dinner table. When autumn frosts reduce the availability of herbaceous vegetation, the eastern cottontail switches to woody plants for food. It readily consumes the bark, twigs, and/or buds of raspberry, apple, red maple, honey locust, staghorn sumac, black cherry, and dozens of other species.

Before giving birth, a female prepares a nest in a natural depression or in a shallow hole that she digs herself and then lines with grass and with fur plucked from her body. Gestation averages 28 days. Although litter size averages 5, it varies from 3 to 8. The altricial newborn weighs 25–30 g (1 oz) and grows at the rate of 2.5 g/day (0.1 oz/day), despite receiving little attention from the mother, who visits the nest only once or twice every 24 hours. The young opens its eyes toward the end of the first week and is fully furred by the end of the second week, when it begins to venture out of the nest. Weaning takes place about one week later. Mating occurs from late March into September, and many adult females commonly produce three litters each year. Eastern cottontails become sexually mature at an age of 2–3 months; consequently, about half the youngsters reproduce in the summer of their birth.

Some eastern cottontails live up to four years in the wild, but the average life span is less than one year. Predators are numerous and include coyotes, foxes, domestic dogs, hawks, and owls. Winter is particularly difficult for cottontails along the northern edge of their range, where persistent snow cover makes the brownish animal highly visible to predators and simultaneously makes escape more difficult. The cottontail is a favorite target of hunters and probably the most important game animal in the southern Great Lakes basin. In addition, some cottontails suffer from tularemia, a bacterial disease that humans occasionally contract while handling an infected carcass; a diseased animal appears sluggish, and its liver and spleen are covered with white spots.

Suggested References: Chapman, Hockman, and Edwards 1982; Chapman, Hockman, and Ojeda 1980; Keith and Bloomer 1993; Woolf, Shoemaker, and Cooper 1993.

Appalachian Cottontail *Sylvilagus obscurus*

Measurements: Total length: 386–430 mm (15–17 in); tail length: 22–65 mm (0.9–2.6 in); hindfoot length: 87–96 mm (3.4–3.8 in); ear height: 54–63 mm (2.1–2.5 in); weight: 0.8–1.0 kg (1.8–2.2 lb).

Appalachian cottontail

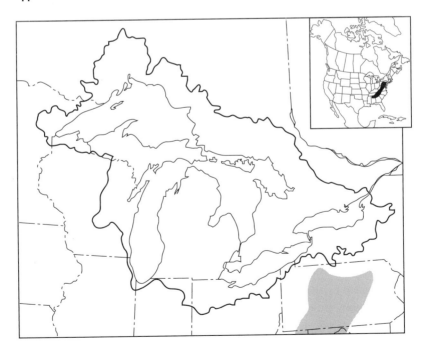

Description: Dorsal fur is yellowish brown washed with black, except for a reddish brown patch over the neck. The animal is lighter in color on the sides and white along the belly; its short, fluffy tail is dark above and white below. This rabbit is quite similar to the more common eastern cottontail. However, most Appalachian cottontails have an indistinct black spot between the ears and lack the white forehead blaze seen on many (but not all) eastern cottontails. In the field, one can easily distinguish this species from hares by the cottontail's small size and rusty neck patch.

Natural History: This cottontail lives along the Appalachian Mountains, from Georgia and Alabama northward to Pennsylvania, where the species most closely approaches the Great Lakes drainage. Within the mountains, this rabbit is restricted to high-elevation habitats that are reminiscent of boreal forests. The Appalachian cottontail frequents areas containing coniferous trees and dense cover provided by shrubby plants of the heath family, such as rhododendron, blueberry, and laurel. Its numbers have steadily declined over the past few decades, perhaps as a result of competition with introduced populations of the eastern cottontail. Biologists long considered the Appalachian cottontail to be just a southern form of the New England cottontail, *Sylvilagus transitionalis,* but differences in chromosomal number and cranial anatomy suggest that the two are truly distinct species.

The diet of the Appalachian cottontail has not been studied in detail, but it probably is similar to that of the eastern and New England cottontails. Presumably, the summer diet consists of mostly green vegetation, such as grass and clover, while the remainder comes from leaves, twigs, and fruits taken from a variety of herbaceous and shrubby plants. In winter, this cottontail probably nibbles on the bark and buds of birch, red maple, aspen, chokecherry, and black cherry; shrubby willows and alders, as well as blackberry and blueberry bushes, round out the winter menu.

In late winter, males come into breeding condition in response to lengthening daylight. However, onset of actual mating activity is less predictable and apparently correlated with temperature; copulation does not occur until fair weather arrives in spring and then continues into September. Many female mammals release an egg from the ovary in response to periodic changes in hormone levels, but all cottontails are "induced ovulators," discharging an egg only after the physical stimulation of mating.

Prior to parturition, expectant mothers build a nest of fur and grass in a shallow depression, about 10 cm (4 in) deep and 13 cm (5 in) wide. Litter size is smaller than in the eastern cottontail and averages 3–4. Gestation probably lasts 28 days. A female leaves her young unattended for prolonged periods, but before departing, she camouflages the nest and young with a layer of fur and grass and then another of leaves and twigs. Newborn rabbits can not maintain a high body temperature, and this covering may keep them warm as well as hide them from predators. Growth is rapid, and lactation lasts only 16 days. Although young males never mate until the following spring, one in five females reproduces during the summer of her birth. Adults commonly rear three litters each year.

Potential predators include owls, hawks, dogs, and foxes. Hunters kill many, but exact numbers are difficult to obtain, because hunters and government agencies frequently do not distinguish between the eastern cottontail and the Appalachian cottontail. Average life span is probaby less than a year.

Suggested References: Chapman et al. 1992; Chapman, Hockman, and Edwards 1982; Merritt 1987.

Snowshoe Hare   *Lepus americanus*

Measurements: Total length: 380–505 mm (15–20 in); tail length: 25–45 mm (1.0–1.8 in); hindfoot length: 120–150 mm (4.7–5.9 in); ear height: 60–70 mm (2.4–2.8 in); weight: 1.4–2.3 kg (3.1–5.1 lb).

Description: In summer, the snowshoe hare has a yellowish brown to rusty brown coat with an indistinct dark band running down the dorsal midline. Belly fur is white, and a cinnamon-brown patch covers the throat. In winter, the animal appears white, except for the black ear tips. Compared to cottontails, the snowshoe hare lacks a rusty brown nape and generally is larger, especially its feet. Compared to other hares, the snowshoe is smaller in all dimensions. Furthermore, the base of each dorsal hair is a dark, slate gray in the snowshoe but white or light gray in other hares.

Snowshoe hare

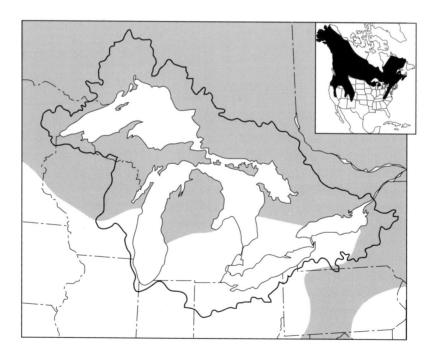

Natural History: The snowshoe hare's domain includes most of Canada and a portion of the northern United States. Unlike the eastern cottontail, the snowshoe hare prefers heavily forested areas with a dense understory and avoids habitats dominated by humans. It thrives in coniferous and mixed woods, is less common in northern hardwood forests, and is particularly at home in low-lying situations such as cedar bogs and spruce swamps. Population density varies in a cyclical manner, with peak densities occurring every 10–11 years. Strangely enough, this population cycle corresponds to a multiyear cycle in sunspot number, and a recent study suggests that there is a link between solar activity, climate, and hare numbers.

The snowshoe hare is well adapted to life in the snowy north. The large hindfeet, densely coated with stiff hairs, actually function as "snowshoes," allowing the animal to walk across deep snow without sinking. Beginning in September, this mammal loses its brown summer fur and acquires the familiar white coat. The transition (molt) from summer to winter pelage is triggered by decreasing daylength and requires 70–75 days for completion. The thick winter coat provides excellent insulation, and the white hairs make the animal extremely difficult to detect against a snowy background. The spring molt begins in early March and is not complete until late May.

This hare remains in dense cover during the day and ventures into forest clearings for feeding only at night. Summer foods include the usual leporid favorites, grasses and clovers, along with dandelion, aster, strawberry, ferns, and even horsetails; it also eats the young leaves of aspen, willow, and birch. Winter foods include the bark, twigs, and buds of woody plants, such as maple, willow, poplar, and hazelnut, as well as the needles of most conifers. Unlike our other leporids, the snowshoe hare is a meat eater, occasionally scavenging from the carcass of a deer or another hare. It forages over a home range that averages 8 ha (20 acres) in extent.

Mating takes place anytime between March and September and is preceded by an energetic courtship involving many jumps and chases and one animal urinating on the other. The young develop inside the uterus for 37 days and weigh 65–80 g (2–3 oz) at birth. Neonates already have a black ear-tip and are fully furred with eyes open. The mother provides no nest and minimal maternal care. She hides the young in a simple form beneath a log, shrub, or other suitable cover, and rests by herself as much as 250 m (800 ft) away. Litter mates separate soon after birth, hide in different locations, and reassemble once each day for feeding; they suckle for 5–10 minutes from the visiting parent before the whole family splits apart again. Weaning generally occurs after four weeks, yet the young hares, or leverets, do not breed until the next spring. Average litter size is four, and a female hare gives birth up to four times each year.

Mortality is high, and no more than 15% of the adults survive until the next year. The snowshoe hare is a source of food for the coyote, red fox, gray wolf, bobcat, lynx, and mink, and humans avidly hunt this hare for its fur and meat. Large owls, such as the great horned owl, are common avian predators. Unlike cottontails, hares rarely suffer from tularemia.

Suggested References: Bittner and Rongstad 1982; Keith and Bloomer 1993; Sinclair et al. 1993.

European Hare   *Lepus europaeus*

Measurements: Total length: 640–700 mm (25–28 in); tail length: 70–100 mm (2.8–3.9 in); hindfoot length: 145–160 mm (5.7–6.3 in); ear height: 79–100 mm (3.1–3.9 in); weight: 3.0–5.6 kg (6.7–12.5 lb).

Description: Throughout the year, fur along the back is rusty brown to yellowish brown with a grizzled look. The European hare is grayish on the sides and rump, white on the chin and belly, and yellowish brown on the throat. Ear tips and top of the tail are black. Large size and chunky build easily separate the European hare from the cottontails. Compared to the white-tailed jackrabbit or snowshoe hare, the European hare has dorsal hairs that are white, and not gray, at their base.

Natural History: The European hare is an introduced species native to the Old World. The animals living in Canada today probably are the descendants of nine individuals that escaped from captivity near Brantford, Ontario, in 1912. The species spread throughout southeastern Ontario and may have crossed the St. Clair River into Michigan by the 1930s. However, it never became established in Michigan, and any specimens found there today likely are recent escapees. This hare was once widespread in New York and adjacent states, but its numbers and range have dwindled considerably; those that remain may be descen-

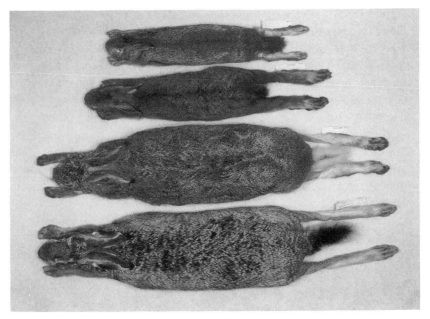

From bottom to top: museum specimens of European hare, white-tailed jackrabbit, snowshoe hare, and eastern cottontail. Range map is for European hare. (Photo by Rod Foster.)

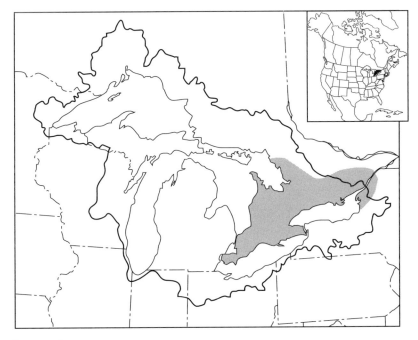

European hare

dants of hares released into the wild in New York between 1893 and 1911.

This species prefers open expanses of cultivated fields, meadows, and pastures. In southern Ontario, where the species is well established, average density is 10 hares/km^2 (25/mi^2). It is nocturnal and spends the day partly concealed in a shallow depression, or form, beneath tall grass or some brush. When hounded, it stays in the open and avoids dense thickets, unlike the smaller cottontails or snowshoe hare that enter tangled areas to elude a predator. The European hare prefers to outrun its pursuers, bounding along at speeds up to 56 km/hour (35 mi/hour). Home range encompasses about 5 ha (12 acres).

The diet is probably similar to that of rabbits and other hares in the Great Lakes region. In summer, it relies primarily on grasses and other herbaceous vegetation, and, in winter, it consumes the bark and twigs of trees and shrubs. During the daylight rest period, the European hare ingests soft greenish fecal pellets. These pellets contain nutrients that were released by bacterial action in the cecum and are a valuable supplement to the diet of all lagomorphs.

The breeding season is characterized by marked aggression between males, who fight and chase each other even during daytime. Mating in the European hare begins as early as December but continues only until midsummer. Gestation lasts about six weeks. The average litter contains four leverets (young hares) that are covered with long, silky hair, have their eyes open, and weigh 130 g (4.5 oz) each at birth. Soon after parturition, the doe separates litter mates and places them into different nearby forms. She then visits these on a rotating basis and nurses each young. Adult females reproduce twice yearly, but newborn females do not become sexually mature until after their first winter.

Fox and coyote are probably common predators throughout the North American range of the European hare, and wild cats, such as bobcat and lynx, may also feast on this large leporid in more secluded areas. Humans actively hunt the European hare for its tasty flesh and because it can be a pest in orchards. This hare has a penchant for the bark of young apple trees, and it may girdle these trees during winter when herbaceous food is scarce.

Suggested References: Dean and de Vos 1965; Schneider 1990.

White-tailed Jackrabbit *Lepus townsendii*

Measurements: Total length: 560–650 mm (22–26 in); tail length: 70–100 mm (2.8–3.9 in); hindfoot length: 135–160 mm (5.3–6.3 in); ear height: 95–110 mm (3.7–4.3 in); weight: 2.5–4.5 kg (5.5–9.9 lb).

Description: In summer, the white-tailed jackrabbit is grayish brown on top, white or pale gray underneath, and has a darker throat. In winter, the animal dons a white coat. The ears are quite long, similar to those of a jackass (hence, the common name), and the tips remain blackish year-round. Longer ears and much longer tail distinguish the jackrabbit from the snowshoe hare and cottontails. In addition, the base of each dorsal hair is light gray in the jackrabbit, compared to dark slate gray in the snowshoe and pure white in the European hare. The tail of the jackrabbit is grayish white all around or with a thin, dark middorsal stripe; the tail of a European hare, in contrast, is glossy black over the entire upper surface.

Natural History: The white-tailed jackrabbit is a denizen of North American grasslands, rarely enters wooded areas, and originally was absent from the heavily forested Great Lakes basin. However, following the conversion of forests into open farmland, this hare naturally spread through Minnesota toward the western shore of Lake Superior and across the Mississippi River into Wisconsin. In addition, humans intentionally released the species into Wisconsin during the early 1900s, hoping to provide another game animal; although once abundant in that state, populations have declined in recent years.

A jackrabbit is not a sociable animal, preferring to rest and forage alone. It is active primarily at night and spends the day in a shallow form dug at the base of a shrub or large rock; such forms are about 50 cm long, 25 cm wide, and up to 20 cm deep (20 × 10 × 8 in). In winter, the jackrabbit shelters in cavities beneath the snow. It generally avoids predators by remaining motionless with ears flat against the body, but if danger approaches too closely, the jackrabbit bounds away in a zigzag fashion. This species eludes enemies by running at speeds up to 55 km/hour (35 mi/hour), covering 5 m (16 ft) with each stride, and it readily enters a pond or stream, swimming with its forefeet only. Its senses of sight, hearing, and olfaction are well developed.

About 75% of the summer diet is grasses and forbs, especially clover,

White-tailed jackrabbit in winter pelage. (Photo by Jerran T. Flinders.)

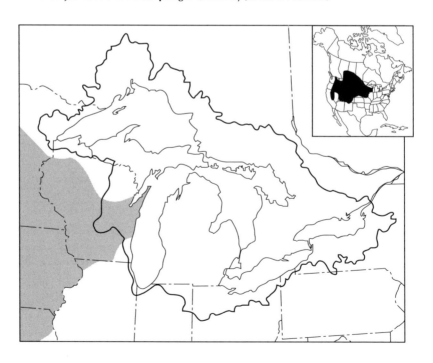

alfalfa, and dandelion. In winter, it feeds primarily on buds and twigs taken from low shrubs. As in all lagomorphs, the jackrabbit gains valuable nutrients by practicing coprophagy.

Courtship is a brief affair, lasting only 5–20 minutes. It occurs during the evening hours and includes a sequence of jumps, circles, and lively chases that precede copulation. Females give birth to a litter of 4 or 5 young, following a 42-day gestation, sometime between April and September. The white-tailed jackrabbit is a synchronous breeder with most females in a population mating and then giving birth at approximately the same time, up to four times each season. This "rabbit" is actually a hare, producing precocial young and providing no nest and no maternal care other than milk. The leverets weigh 90 g (3.2 oz) at birth and begin eating grass after just two weeks, yet they do not become totally independent until about two months of age. Both males and females reach sexual maturity the following spring.

Mammalian predators include the red fox, coyote, lynx, bobcat, and weasels. The great horned owl, various hawks, and even the golden eagle are known to attack from above. In addition, humans take a significant toll through hunting, and automobiles strike many hares that forage on grassy roadsides. Some of these jackrabbits live up to eight years in the wild.

Suggested Reference: Lim 1987.

Rodents
Order Rodentia

Rodents occur naturally on all habitable continents, including Australia, where rodents and bats are the only native eutherian mammals. Worldwide, rodents occupy a variety of habitats, including frigid polar regions, harsh deserts, high mountains, and lush tropical forests. Most species are terrestrial in habit; however, some spend much of their time in trees, and others permanently reside underground. Numerous rodents have evolved webbed feet and flattened tails for swimming, fleshy membranes for gliding, or huge, kangaroolike feet for hopping. Most are small bodied, but they range from the Old World harvest mouse at 5 g (0.2 oz) to the portly capybara at 50 kg (110 lb). This group is the most successful of all mammalian orders, containing 2,022 living species, or about 40% of all mammals. Not surprisingly, rodents also dominate the mammalian fauna of the Great Lakes basin, where there are 30 species, representing 21 genera and 6 families.

Distinctive traits of all rodents are a single pair of curved incisors, on both the upper and lower jaw, and the separation of the incisors from the cheek teeth by a wide space or diastema (fig. 8). The incisors grow throughout life, so that the surface worn away by constant gnawing is gradually replaced. The front surface of the incisors is made of hard enamel (usually orange or yellow in color), whereas the rear portion of the tooth consists of softer dentine; differential wear between the two surfaces maintains a sharp, chisellike edge. In some species, the cheek teeth have low, rounded cusps for crushing, but many other rodents have distinct ridges of enamel for grinding food. The bones and muscles of a rodent jaw are arranged in such a way that the animal either gnaws with its incisors or chews with its cheek teeth; it cannot do both at the same time.

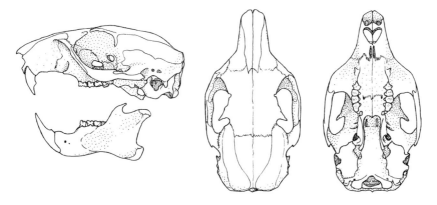

Fig. 8. Skull of an eastern fox squirrel, family Sciuridae

Squirrels
Family Sciuridae

Squirrels first appear in the fossil record in rocks from the Oligocene epoch that are over 30 million years old. Today, the family lives in most parts of the world except southern South America, portions of the Middle East, and the Australian region. Taxonomists classify modern squirrels into 273 species, 10 of which live in the Great Lakes area.

The ability of some squirrels to thrive in dense urban habitats makes this family the most familiar of all wild rodents. Although most people picture a squirrel as a bushy-tailed animal scampering along a leafy branch, this family actually has a variety of life-styles. Besides tree squirrels, there are nocturnal flying squirrels that glide gracefully from tree to tree and short-tailed ground squirrels, such as the prairie dog, that make their home in subterranean burrows. Many squirrels hibernate in winter, and some estivate in summer. They feed largely on vegetation, nuts, and seeds, supplemented with an occasional insect, small verte-brate, or bird egg.

The skull of a squirrel is easy to distinguish from that of other Great Lakes rodents. All squirrels have a well-developed postorbital process; this is a slender, often sharply pointed projection from the frontal bone, on top of the skull, just behind each eye (see fig. 8). In addition to the obvious incisors, squirrels have four cheek teeth on either side of the lower jaw and four or five cheek teeth on each side of the upper jaw, for a total of 20 or 22 teeth. Members of the Geomyidae (pocket gopher),

Castoridae (beaver), and Erethizontidae (porcupine) also have 20 teeth, but these families lack the obvious postorbital process.

Least Chipmunk *Tamias minimus*

Measurements: Total length: 185–222 mm (7.3–8.7 in); tail length: 80–100 mm (3.1–3.9 in); hindfoot length: 28–35 mm (1.1–1.4 in); ear height: 13–18 mm (0.5–0.7 in); weight: 42–53 g (1.5–1.9 oz).

Description: The least chipmunk is the smallest squirrel in the Great Lakes basin. Along the back, it has five dark brown stripes running from the neck to the base of the tail. Fur is white between the lateral pairs of stripes, but light brown adjacent to the median stripe. The sides appear orangish brown, and the belly is grayish white. It can only be confused with the eastern chipmunk, which is larger, has a distinct reddish brown patch on the rump, and stripes that do not extend to the base of the tail. Also, a least chipmunk has five upper cheek teeth on each side, compared to only four in the eastern chipmunk.

Natural History: The least chipmunk dwells in and about the boreal forests of central and western North America. It avoids the closed interior of mature climax stands of conifers, preferring the more open forest edge and internal forest clearings. Disturbed areas surrounded by open stands of aspen and pine often harbor large populations. The least and eastern chipmunks occasionally occur together in deciduous habitats, but the eastern chipmunk prefers areas with greater cover. Typical population density is 1–6/ha (0.5–2.5/acre).

A least chipmunk is territorial, defending its nest site against intruders, but direct confrontations with neighbors occur less often than in the eastern chipmunk. Although it commonly scurries along the ground, this squirrel also forages in shrubs and trees, at heights up to 9 m (30 ft). The least chipmunk ceases all activity prior to nightfall and remains in its nest until after sunrise. This small squirrel is active only from April until October; with the onset of cold weather, it retires to an underground burrow, frequently enters torpor, and lives off stored food until spring.

It builds its winter nest on top of a food cache in a chamber up to 1 m (3 ft) below the surface. The nest consists of dried grass or shredded bark lined with feathers, fur, or soft vegetable matter. The single surface

Least chipmunk. (Photo by Gilbert L. Twiest.)

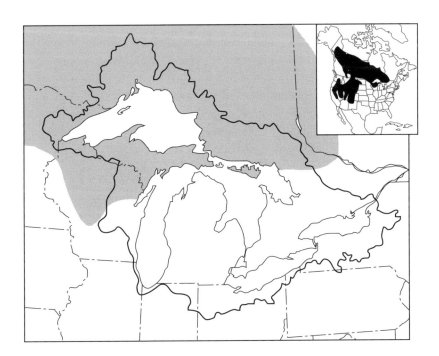

entrance to the burrow has no telltale pile of dirt nearby, because the chipmunk removed excavated soil through a "work hole" that it plugged at the end of construction. In summer, a least chipmunk often abandons its burrow and builds another leafy nest in a rotting log or tree cavity.

Nuts, fruits, and seeds are preferred foods. Although it relishes blueberries, raspberries, and the like, it does not eat the pulp of these fruits. Instead, the squirrel opens the berries, removes the seeds, and leaves behind a pile of soft tissue. A least chipmunk has distensible cheek pouches that open inside its mouth; when foraging, it fills the pouches with seeds and nuts, carries them back to its burrow, and stores them for future use. One enterprising chipmunk managed to accumulate almost 500 acorns and 2,700 cherry pits. This rodent obtains extra protein in the summer by preying on insects and consuming an occasional bird egg.

The testes probably reach maximum size in April, and most mating occurs at that time. A least chipmunk bears a single litter each year, although a female that loses her first litter to predators may try again. Most births occur in late May and early June, and an average litter contains 4–7 naked young weighing about 2.2 g (0.08 oz) each. Gestation takes about 30 days, and lactation lasts another 60 days or less. Youngsters do not breed until the following year, when they are about 10 months old.

Predators include snakes, hawks, weasels, and foxes, as well as domestic dogs and cats. There are no data on longevity.

Suggested References: Banfield 1974; Forbes 1966.

Eastern Chipmunk *Tamias striatus*

Measurements: Total length: 225–266 mm (8.9–10.5 in); tail length: 65–110 mm (2.6–4.3 in); hindfoot length: 32–40 mm (1.3–1.6 in); ear height: 12–20 mm (0.5–0.8 in); weight: 66–115 g (2.3–4.1 oz).

Description: The back of an eastern chipmunk is grayish brown to reddish brown with an obvious red patch on the rump. Underparts are white. There are five dark stripes along the back and sides, the longest of which runs down the midline. Between the paired dark lateral stripes is a narrow band of white. The stripes distinguish the eastern chipmunk from all other rodents, except in the northern Great Lakes region, where the eastern and

Eastern chipmunk

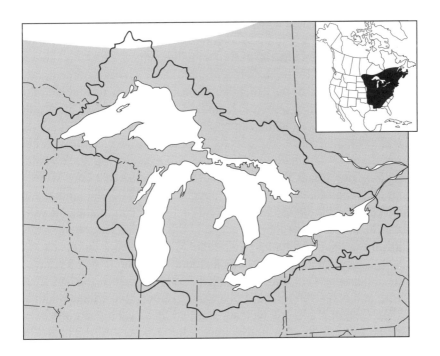

least chipmunks occur together. However, the stripes of the least chipmunk extend to the base of the tail, whereas they do not cross the rump in the larger-bodied eastern chipmunk.

Natural History: This ground-dwelling mammal is the common chipmunk of eastern North America. It inhabits open deciduous forests where stumps, logs, or rocky outcrops provide ready cover. Mature beech-maple forest is perhaps the ultimate habitat for the eastern chipmunk, but it also lives in brushy areas and, to a lesser extent, coniferous forests. This squirrel avoids swampy sites. The eastern chipmunk is somewhat tolerant of humans and occasionally burrows under buildings or stone foundations in rural areas. Typical density is 10–22/ha (4–9/acre).

For a burrow, it excavates a series of interconnected galleries up to 10 m (33 ft) in length and less than 1 m (3 ft) below the surface. The chipmunk uses one room as a nest site, assembling crushed or chewed leaves into a soft bed about 30 cm (1 ft) in diameter. Other chambers are used to store food or debris. A surface entrance measures 5 cm (2 in) across and appears quite tidy because the squirrel does not leave excavated soil in a conspicuous pile but spreads it among the surroundings or uses it to plug an older entrance.

An eastern chipmunk is busiest during midmorning and midafternoon. It is solitary and territorial, particularly in the immediate vicinity of its burrow. A resident visually and vocally threatens an intruding chipmunk before engaging in a vigorous and often noisy chase. So intent are the squirrels on their dispute that the chase often passes unabated within a few paces of startled humans. Although this species forages mostly along the ground, it occasionally climbs a tree or shrub.

Dietary staples are fruits, seeds, and nuts, supplemented with mushrooms, insects, earthworms, slugs, and bird eggs. A chipmunk transports dry food items back to its burrow in cheek pouches that open within the mouth. When totally filled, each pouch is as large as the animal's head and contains hundreds of seeds. Food caching occurs at any time of year, but the animal is particularly active in the autumn, storing food for the upcoming winter. Although a chipmunk saves energy during winter by periodically entering torpor, this mammal lacks the large fat stores of a true hibernator. Consequently, it must arouse frequently, feed off hoarded food, and may even forage above ground during mild winter weather.

An eastern chipmunk typically has two litters each year, one in early spring and the other in midsummer. After a gestation of 31 days, 2–5 blind and hairless young are born in an underground nest. A baby chip-munk weighs 3 g (0.1 oz) at birth and reaches 35 g (1.2 oz) before it is weaned, 40 days later. Most young do not breed until the spring follow-ing their birth.

An eastern chipmunk makes an excellent meal for most diurnal preda-tors. The long-tailed weasel, ermine, lynx, bobcat, red fox, and coyote are consistent mammalian predators, along with the domestic dog and cat. Large raptors, such as the red-tailed hawk and goshawk, also feed on this squirrel. Although a few chipmunks live up to eight years in nature, most survive less than two years.

Suggested Reference: Snyder 1982.

Woodchuck or Groundhog   *Marmota monax*

Measurements: Total length: 530–650 mm (21–26 in); tail length: 100–160 mm (3.9–6.3 in); hindfoot length: 65–95 mm (2.6–3.7 in); ear height: 25–40 mm (1.0–1.6 in); weight: 2.3–5.0 kg (5–11 lb).

Description: The woodchuck is, by far, the largest squirrel in the Great Lakes area. It has a chunky body set atop short, powerful legs that are well adapted for digging. The dark-colored tail is short compared to other squirrels and only equals about 25% of the total length. Dorsal guard hairs are yellowish to reddish brown and tipped with white, giving the animal a frosted appearance. Ears are low and rounded, and the feet are black. The incisors of many woodchucks lack the yellowish pigmentation characteristic of other rodents.

Natural History: The woodchuck ranges from Alabama to Labrador to Alaska. Originally this ground squirrel lived in open forests but it has adapted well to human intrusion. Today, rolling farmland interspersed with grassy pastures, small woodlots, and brushy fencelines is typical woodchuck habitat.

Here, it excavates a comfortable burrow in well-drained soil, often on a natural hillside or highway embankment. A mound of fresh dirt surrounds a burrow entrance that measures 25–30 cm (10–12 in) in diameter and leads to a tunnel that usually slopes upward to prevent rainwater from entering. In addition, the solitary resident often digs a

Woodchuck

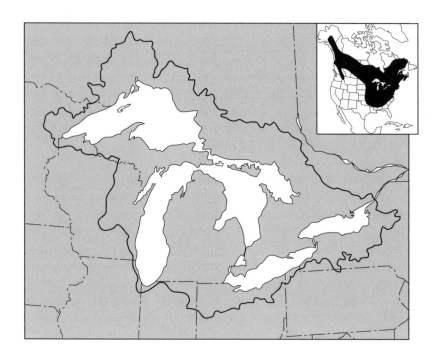

slightly smaller "plunge hole" that is not surrounded by a conspicuous mound and drops 60 cm (2 ft) straight into the subterranean system. Tunnels extend 6–9 m (20–30 ft) underground and include a slightly enlarged nest chamber filled with dry leaves. A woodchuck often uses more than one burrow during a single season.

This diurnal animal leaves to forage along well-defined paths radiating out from the burrow entrance. The woodchuck feeds mostly on grasses and herbaceous plants, such as dandelion, daisy, goldenrod, and clover, but this ground squirrel also competes with farmers and gardeners for carrots, beans, celery, soybeans, and corn. In early spring, before green vegetation is available, a woodchuck eats the bark, buds, and twigs of dogwood, sumac, and black cherry. Although seemingly out of its element, this ungainly creature occasionally climbs trees in search of food.

In contrast to the nuts and seeds eaten by chipmunks, the herbs and grasses preferred by a woodchuck do not store well, and a woodchuck does not cache food for winter use. Instead it is a true hibernator that deposits extra body fat in preparation for a long winter dormancy, during which the squirrel maintains a body temperature of only 4°C (40°F), except for brief periodic arousals. Hibernation generally begins in October and ends in March or April; contrary to popular belief, all woodchucks are still in deep hibernation on Groundhog Day (2 February). A woodchuck loses 30% of its body mass over winter as it utilizes its excess fat.

In the spring, males in breeding condition venture aboveground first and fight vigorously with each other. Females remain in hibernation for an extra 1–3 weeks and are mated almost immediately after they emerge. The single annual litter of 2–5 young is born in April or May, 32 days after copulation. Newborn woodchucks weigh 26 g (0.9 oz). Body hair appears after two weeks, and eyes open after four weeks; the mother weans her offspring by the end of the sixth week, and most young quickly disperse. Although a newborn woodchuck becomes sexually mature the following spring, it does not reach adult size until the end of its second summer.

Some woodchucks survive for six years in the wild, but most do not see their third summer. A large body size protects a woodchuck from most predators, but some, especially young ones, are taken by red fox, gray fox, black bear, lynx, bobcat, and various hawks. Humans hunt the woodchuck for food, exterminate it as an agricultural pest, and accidentally flatten thousands each year on the highway. When frightened, a woodchuck emits a shrill whistle that earns it the name "whistle pig" in some areas.

Suggested References: Lee and Funderburg 1982; Swihart 1992.

Franklin's Ground Squirrel   *Spermophilus franklinii*

Measurements: Total length: 350–420 mm (14–17 in); tail length: 125–160 mm (4.9–6.3 in); hindfoot length: 50–57 mm (2.0–2.2 in); ear height: 16–18 mm (0.6–0.7 in); weight: 370–500 g (13.1–17.6 oz).

Description: From nape to rump, Franklin's ground squirrel appears a dark yellowish brown, sprinkled with black. Ventral fur is thinner and lighter, often a yellowish white. At a distance, this species resembles a small gray squirrel, but Franklin's ground squirrel has smaller and more rounded ears and a shorter and less bushy tail. The tail of this ground squirrel amounts to less than 40% of the animal's total length.

Natural History: Franklin's ground squirrel is primarily a resident of tall grass prairies, and it occurs only along the western fringe of the Great Lakes drainage. It prefers grass of intermediate height, tall enough to obscure the animal when on all four feet, but low enough to see over when standing upright on the hind legs. It often lives along forest-grassland edges and in grasses bordering railroads and highways. These squirrels tend to live in loosely associated groups, rather than spreading evenly throughout appropriate habitat.

A Franklin's ground squirrel excavates a burrow in well-drained soil, often into the side of a hill, ditch, or railroad embankment. A mound of soil marks the main entrance, but it is often concealed by tall grass. The main tunnel, 8–10 cm (3–4 in) across, branches frequently and descends up to 2 m (7 ft) below the surface. A side passage contains a nest made from dried plant material.

This animal is fairly sedentary and rarely ventures more than 100 m (330 ft) from its burrow in search of food. It occasionally climbs a tree but forages mostly along the ground during daylight hours. The Franklin's ground squirrel eats green vegetation as well as fruits and seeds from a variety of plants, including grass, thistle, dandelion, clover, and blackberry, along with cultivated grains and garden vegetables. Up to one-

Franklin's ground squirrel

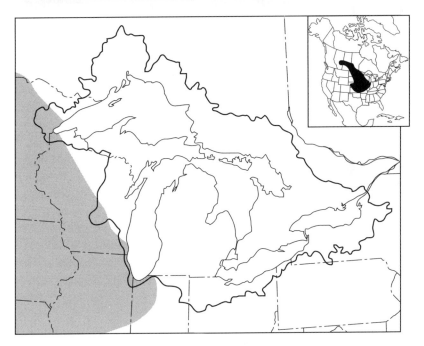

third of its diet consists of animal flesh, especially insects. Caterpillars, grasshoppers, crickets, and ants are the most common prey, but occasionally this rodent consumes a frog, toad, nestling bird, bird egg, mouse, or young rabbit.

The Franklin's ground squirrel spends 7–8 months each year hibernating in its underground burrow. Adults enter hibernation first, often as early as September, but the inexperienced young-of-the-year need an extra 1–4 weeks of foraging before they are fat enough to survive the winter. Males leave hibernation beginning in late March, and females follow within a week or so.

Mating takes place soon after females emerge, and they give birth about 26–28 days later. Average litter size is 7–8 but varies from 4 to 11. As in other squirrels, newborn are altricial. Fur gradually appears over the first two weeks of life, eyes finally open by 20 days of age, and young leave the natal burrow for good after 40 days. Adults produce only one litter each year, and youngsters become sexually mature the spring following their birth.

Red-tailed hawks are a consistent threat when this squirrel is above ground, and badgers occasionally dig it from its burrow. Foxes, coyotes, skunks, and weasels also prey on Franklin's ground squirrel. As in many hibernating species, failure to store sufficient fat for the winter probably is an important mortality factor for young-of-the-year. Populations seem to undergo large fluctuations every 4–6 years for no apparent reason.

Suggested References: Choromanski-Norris, Fritzell, and Sargeant 1986 and 1989; Iverson and Turner 1972; Turner, Iverson, and Severson 1976.

Thirteen-lined Ground Squirrel
Spermophilus tridecemlineatus

Measurements: Total length: 225–300 mm (8.9–11.8 in); tail length: 75–109 mm (3.0–4.3 in); hindfoot length: 32–41 mm (1.3–1.6 in); ear height: 7–11 mm (0.3–0.4 in); weight: 100–220 g (3.5–7.8 oz).

Description: The thirteen-lined ground squirrel has a slender body, bristly hair, thin tail, and very short ears. Along the back and sides, it has seven dark brown stripes alternating with six longitudinal bands that are gray to tan in color. Within each dark stripe is a series of tan rectangles.

Thirteen-lined ground squirrel

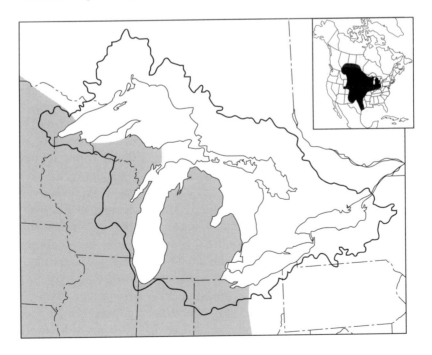

Chipmunks are generally smaller and not as elaborately striped as the thirteen-lined ground squirrel.

Natural History: During the past two centuries, this prairie mammal slowly expanded its range northward and eastward, into the Great Lakes region, as forests were cut and swamps were drained to make way for European settlers. Unlike Franklin's ground squirrel, this species prefers short grass that it can easily see over without standing up. Consequently, the thirteen-lined ground squirrel is at home in open areas such as mowed lawns, manicured golf courses, and well-grazed pastures. It occasionally lives in weedy areas adjacent to highways or railroads, and it always avoids the woods. Like Franklin's ground squirrel, this species is semicolonial.

The thirteen-lined ground squirrel is completely diurnal and most active around midday. It prefers warm, sunny days, and during poor weather, it remains underground, subsisting on stored food. It digs a number of short, shallow, often blind-ending burrows for emergency use and a deeper more complex system for nesting or hibernation. Home range averages 4.7 ha (11.6 acres) for males and 1.4 ha (3.5 acres) for females.

Vegetation dominates the diet, but this rodent is omnivorous. Caterpillars, grasshoppers, and beetles are common animal food, although it also takes small vertebrates and bird eggs and readily consumes carrion. It eats the leaves of grass and clover, but prefers seeds; in fact, the name *Spermophilus* means "seed lover." The thirteen-lined ground squirrel relishes seeds from crabgrass, sunflower, chickweed, thistle, ragweed, and vetch, as well as corn, wheat, and oats. Many seeds are carried back to the burrow in bulging cheek pouches and cached for later use.

In the autumn, this mammal rapidly gains weight, storing up to 4 g (0.1 oz) of fat each day, in preparation for winter dormancy. When hibernating, body temperature drops close to burrow temperature, and heart rate decreases from more than 200 beats/minute to less than 20 beats/minute. Hibernation begins in September and ends in late March or April. As in other ground squirrels, juveniles are the last to disappear underground in autumn, and males arouse before females in spring.

Thirteen-lined ground squirrels are promiscuous; they mate within five days of spring emergence, and copulation serves as the stimulus for ovulation, which occurs 1–2 days later. After a gestation of 28 days, a female gives birth to a single annual litter that averages 8–9 offspring.

The blind and hairless newborn grow at a rate of 1.3 g/day (0.05 oz/day), and they leave the burrow for the first time 28 days after birth. The family breaks up 8–14 days later, at which time youngsters set up housekeeping on their own, usually within 100 m (330 ft) of where they were born.

The large litter size correlates with high mortality. Up to 90% of newborn die from predation before hibernation begins. Badgers dig up the squirrel's burrow, weasels and snakes venture underground looking for them, and hawks pick off the young when they dare to come aboveground. Foxes, coyotes, house cats, and skunks also feast on the thirteen-lined ground squirrel.

Suggested References: Foltz and Schwagmeyer 1989; Murie and Michener 1984; Streubel and Fitzgerald 1978.

Eastern Gray Squirrel *Sciurus carolinensis*

Measurements: Total length: 425–500 mm (17–20 in); tail length: 180–250 mm (7.1–9.8 in); hindfoot length: 60–74 mm (2.4–2.9 in); ear height: 25–35 mm (1.0–1.4 in); weight: 350–700 g (12–25 oz).

Description: In most gray squirrels, the body and tail have a distinct grayish cast caused by silvery tips on many of the hairs. Underparts are yellowish white, and the bushy tail is as long as the body alone. There is a white eye ring and a white patch behind each ear, especially in winter specimens. One might confuse this species with the heavier fox squirrel, which lacks silver-tipped hairs, or the smaller Franklin's ground squirrel, which also lacks the silver and has a shorter tail and smaller ears. Some gray squirrels appear totally black, and this melanistic phase is common in various parts of the Great Lakes region; this is the species that people are most likely to refer to as the "black" squirrel.

Natural History: The gray squirrel lives throughout the East, from the Gulf Coast to the northern Great Lakes. Look for this squirrel wherever deciduous trees are abundant—in extensive forested tracts, dense woodlots, riparian strips, or even residential neighborhoods and city parks. It avoids young, second-growth woods in favor of large, mature

Eastern gray squirrel. Typical coloration (left), black color phase (right). (Photos from the Roger W. Barbour Collection, Morehead State University, and by Gilbert L. Twiest, respectively.)

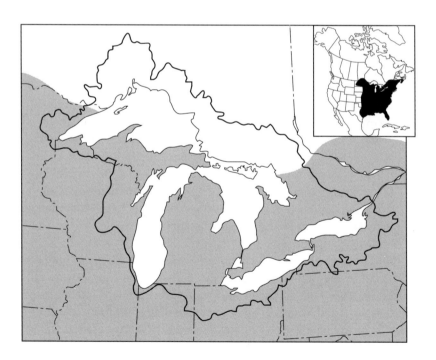

trees that provide nesting hollows and edible nuts or seeds. Sizeable stands of walnut, hickory, maple, or beech are preferable, although it also frequents mixed deciduous/coniferous stands.

In the summer, most foraging takes place just after dawn and before dusk, but in the winter, a gray squirrel is busier during the warmer mid-day hours. It is active year-round, never hibernating, although it may remain in the nest during cold spells. This squirrel spends most of its time in trees, using its long tail as a counterweight to maintain balance as it leaps from branch to branch. It leads a solitary life and rarely strays more than 100 m (330 ft) from its nest tree.

An eastern gray squirrel seeks refuge in either a tree cavity or an exposed leafy nest. To construct an exposed nest, the squirrel cuts numerous twigs, with the leaves still attached, and lodges these into the fork of a substantial tree branch. Although the nest appears haphazard from below, it provides a comfortable, waterproof home for a gray squirrel in the summer. Such a shelter, however, does not provide enough protection against winter cold or for a female raising a brood. During these times, the squirrel uses a tree hollow. An ideal cavity is at least 30 cm (12 in) deep and has an opening 8 cm (3 in) or more in diameter. A gray squirrel prefers a hollow high in a hardwood tree and often uses a cavity originally excavated by a woodpecker. Inside, the rodent arranges a soft, warm nest from shredded bark, leaves, and grass.

This species primarily feeds on acorns, walnuts, hickory nuts, and the like. In late summer, a gray squirrel becomes a "scatter hoarder," burying these edible items in numerous shallow holes, one nut to a hole. Later, the animal relocates each hidden treasure with a keen sense of smell and relies on such items as the main food source in winter. Up to 85% of the buried nuts are retrieved—but not always by the same squirrel that hid them. In spring and summer, it supplements the diet with seeds, mushrooms, buds, flowers, fruits, and insects.

A gray squirrel produces two litters each year after a 44-day gestation; the first litter is born in March or April, and the second appears in July or August. Litters usually contain 2–4 altricial young weighing 15–18 g (0.5–0.6 oz) each. A mother begins to wean her offspring between 50 and 60 days after birth, but they often remain with her until she bears the next litter.

A few have survived for more than 15 years in captivity, and the record in the wild is almost 13 years. The majority, however, live 2 years or less. Major predators are hawks, bobcats, and wild and domestic canids; owls, snakes, and weasels take a few. Millions are killed by hunters or crushed by automobiles every year.

Suggested Reference: Koprowski 1994a.

Eastern Fox Squirrel   *Sciurus niger*

Measurements: Total length: 500–590 mm (20–23 in); tail length: 220–270 mm (8.7–10.6 in); hindfoot length: 62–80 mm (2.4–3.1 in); ear height: 22–32 mm (0.9–1.3 in); weight: 700–1,100 g (25–39 oz).

Description: This tree squirrel is larger than any other squirrel in our area, except the ground-dwelling woodchuck. Dorsal fur is brownish mixed with black, but the face, ears, feet, undersides, and tail are a distinctive dull orange (fulvous). The bushy tail is about as long or longer than the body itself. Although similar in shape to a gray squirrel, the fox squirrel is larger, lacks the silver cast, and is missing the white eye ring and ear patch typical of its gray cousin.

Natural History: Today, the fox squirrel ranges over most of the eastern United States, although it does not occur as far north as the gray squirrel. The fox squirrel prefers deciduous trees in areas that lack a well-developed understory. It frequents open woodlots, forest-field edges, and urban situations. This species avoids the shaded interior of vast forests, and it was rarely found in the Great Lakes drainage before European settlement. However, deforestation has allowed the fox squirrel to spread into our area at the expense of the gray squirrel, which prefers extensive tracts of mature trees. Although suitable habitat exists in southern Ontario, the fox squirrel has not yet conquered the water barrier separating the United States from mainland Ontario. The only fox squirrels in Canada live on Pelee Island, in Lake Erie, where the species was introduced, and in a small area of southern Manitoba.

A fox squirrel starts foraging later in the day than a gray squirrel, stays busy during most of the day, and occasionally is active on moonlit nights as well. It is most often solitary, but two squirrels sometimes share the same winter nest. Its home range is larger than that of a gray squirrel but usually amounts to only 2–4 ha (5–10 acres). This species is not territorial, and home ranges of different individuals often overlap. Perhaps as

Eastern fox squirrel. (Photo by Dianne M. Jedlicka.)

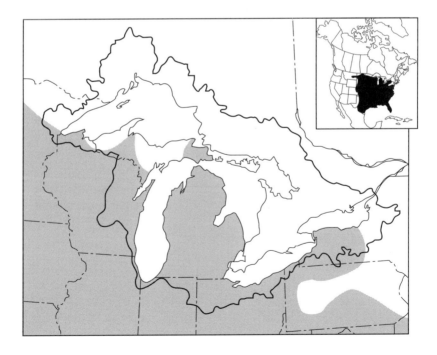

a consequence of its preference for more open habitat, a fox squirrel spends more time on the ground than a gray squirrel.

This large squirrel prefers hollows in deciduous trees for overwintering and raising young. A typical nest cavity measures 15 cm (6 in) across and 35 cm (15 in) deep, has an opening about 8–13 cm (3–5 in) wide, and contains an insulating layer of dry leaves or grass. In the summer, this species often builds a bulky nest of leaves and twigs, high in the branches of an oak, pine, elm, or maple. In southern parts of our area, the fox squirrel occupies these exposed nests throughout winter if suitable tree cavities are not available.

In spring and summer, this rodent eats the buds, flowers, and fruits of woody plants such as maple, elm, and willow. Berries, grapes, and cherries also are eaten when in season, and the fox squirrel apparently pilfers corn more often than the gray squirrel. Extra protein comes in the form of an occasional grub, caterpillar, egg, or young bird. In late summer, the squirrel scatter hoards, burying individual hickory nuts and acorns 2–3 cm (1 in) below the ground. It relies heavily on this hidden food over the winter and relocates up to 99% of the cache using an acute sense of smell.

Although yearling females often produce only a single litter in summer, adult females give birth twice each year, once in March or April and again in July. The gestation period is 44 days, and average litter size is 3–4. The newborn are naked, pink, and blind and weigh about 15 g (0.5 oz) each. Growth is slow. The young do not start taking solid food until 10 weeks after birth and do not become independent for another 3 weeks or more.

An urban fox squirrel from Illinois survived 13 years in the wild, and some have made it to 15 years in captivity. Most, however, do not live for more than a year or two. Many fall to hunters and automobiles each year, and this tree squirrel is common prey for hawks, foxes, coyotes, domestic dogs, and weasels. Winter mortality is highly variable and strongly correlated with mast (nut) production the previous summer.

Suggested Reference: Koprowski 1994b.

Red Squirrel *Tamiasciurus hudsonicus*

Measurements: Total length: 280–345 mm (11–14 in); tail length: 100–145 mm (3.9–5.7 in); hindfoot length: 40–55 mm (1.6–2.2 in); ear height: 18–26 mm (0.7–1.0 in); weight: 135–250 g (4.8–8.8 oz).

Description: In summer, the red squirrel has a dorsal band of pale reddish gray to reddish brown hairs running from its head to the tip of its tail. The body sides appear olive gray and are separated from whitish underparts by a narrow black line. There also is a white ring around the eye. In winter, the dorsal red band is brighter, but the black side stripe is less obvious. The winter coat also includes blackish ear tufts that are missing or poorly developed in summer. The small size and reddish cast distinguish this species from other tree squirrels in the Great Lakes region.

Natural History: This species is the most common tree squirrel in boreal regions. The red squirrel prefers extensive stands of evergreen trees or mixed coniferous/deciduous woodland. It may live in pure deciduous forests, particularly in the southern Great Lakes basin, but even there, it is more likely to take up residence if at least a few conifers grow nearby.

A red squirrel is diurnal, energetic, and extremely vocal. It is asocial and vigorously protests when another squirrel (or human) approaches its den site or favored feeding spot. It initially scolds the intruder with a series of "*tchick-tchick-tchick*" calls, pausing slightly between sounds, and accompanying each with a tail jerk. Further disturbance results in a prolonged "*tcher-r-r-r*," sometimes with a bit of foot stomping added for emphasis. Animals as large as a fox squirrel, and even a blue jay or crow, may be chased after ignoring repeated warnings from an angry red squirrel.

This small tree squirrel is most active near dawn and dusk, except on cold winter days when it restricts activity to midday hours. A tree cavity is the preferred den site, but if one is not available, the squirrel weaves a basketball-sized nest from leaves and twigs high in a tree; inside, the nest chamber is 13 cm (5 in) in diameter and padded with shredded bark, moss, or pine needles. Unlike fox or gray squirrels, the red squirrel occasionally nests on the ground or even underground, especially in the winter. Subterranean burrows measure up to 5 m (16 ft) in length and often contain cached food. In the winter, a red squirrel frequently travels in tunnels beneath deep snow.

Red squirrel

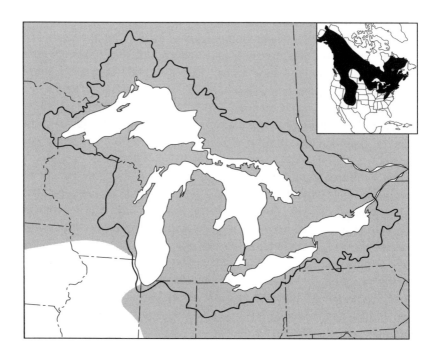

The favorite food is seeds extracted from cones of spruce, fir, larch, and especially hemlock or pine; a large pile of cone scales (a "midden"), often accumulates beneath a lofty feeding perch and indicates the presence of this mammal. In deciduous forests, it eats walnuts, acorns, hazelnuts, and similar foods. During spring and summer, it consumes large quantities of buds, seeds, fruits, mushrooms, and even maple sugar. The squirrel first bites through the bark of a maple tree and then leaves for a day or two; in the meantime, sap oozes down the tree and water slowly evaporates from the sap, leaving behind a sticky treat that the squirrel eagerly consumes upon its return. This species is more carnivorous than gray or fox squirrels and frequently preys on insects, young and adult birds, mice, voles, and young rabbits. In late summer, it caches cones and nuts for winter consumption in large piles stacked next to a stump or hidden in an underground cavity.

Courtship is a brief affair, involving soft vocalizations and energetic chases, that takes place in February or March and again in June or July. A red squirrel living in the far north, however, often forgoes the summer copulation and produces just a single annual litter. After 31–35 days of uterine development, a litter of 3–6 young is born. At birth, each weighs almost 7 g (0.2 oz) and measures 70 mm (3 in) in length. Light fur appears after 10 days, and eyes open after about one month. Mothers wean their brood 7–9 weeks after birth, and the young squirrels soon disperse.

Avian enemies include just about every raptor in our area, ranging from the tiny American kestrel to the magnificent great horned owl. The fisher, marten, bobcat, lynx, mink, coyote, and red fox are common mammalian predators. Trappers and hunters take a few red squirrels, and many are killed on highways. Captive individuals have lived for 10 years.

Suggested References: Lair 1985; Heinrich 1992.

Northern Flying Squirrel *Glaucomys sabrinus*

Measurements: Total length: 245–310 mm (10–12 in); tail length: 110–150 mm (4.3–5.9 in); hindfoot length: 35–40 mm (1.4–1.6 in); ear height: 18–26 mm (0.7–1.0 in); weight: 70–130 g (2.5–4.6 oz).

Description: Our two flying squirrels have a loose fold of skin, the patagium, along each side of the body, from ankle to wrist. When the

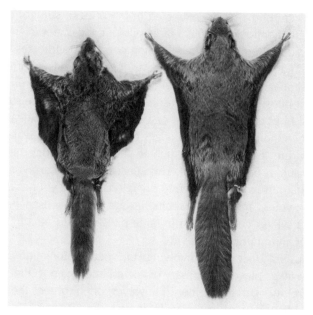

Museum specimens of northern (right) and southern (left) flying squirrels. Range map is for the northern flying squirrel. (Photo by Rod Foster.)

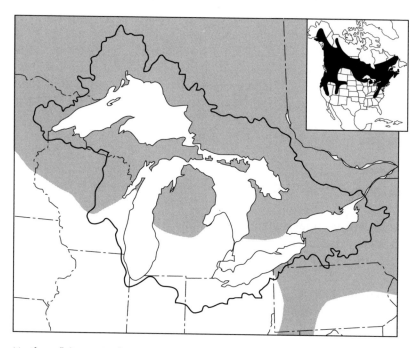

Northern flying squirrel

legs are extended, the patagium forms a kitelike surface that allows the animal to glide from tree to tree; a special cartilaginous support at the wrist extends the membrane farther out than the leg itself. Both species have a silky brown coat, extremely large black eyes for nighttime activity, and a very flattened tail that aids in gliding. However, the northern flying squirrel is much larger than its southern relative. In addition, the northern flying squirrel has belly hairs that are slate gray at the base and white at the tip, whereas those of the southern flying squirrel are completely white.

Natural History: As its name implies, the northern flying squirrel dwells in forests across the northern United States and much of Canada. It is most at home in mixed forests containing mature deciduous and coniferous trees, although it also frequents pure stands of either type. The geographic range of the northern and southern flying squirrels overlap in the Great Lakes basin, and both species occasionally live in the same patch of woods.

Like many squirrels, the northern flying squirrel builds a sheltered nest inside a tree cavity or a spherical nest in the branches of a tree. An outside nest is typically in a conifer, 1–18 m (3–60 ft) above the ground, close to the trunk, and fashioned from twigs, bark, and moss. This squirrel is an opportunist and occasionally uses an old bird's nest as a convenient platform on which to construct its own home. The squirrel softens its nest cavity with feathers, fur, conifer needles, leaves, or finely divided bark.

During the winter, this beautiful mammal shuns exposed nests in favor of a tree hollow. Although a flying squirrel remains in its nest during severe storms, it actively forages during cold weather, even at temperatures below -20°C (-4°F). Up to nine of these small squirrels share the same winter nest and presumably huddle together for added warmth; temperatures inside an occupied nest cavity may be more than 20°C (36°F) above outside air temperature.

The northern flying squirrel is strictly nocturnal, with most activity occurring in the first two hours after sunset and in the hour before sunrise. It forages along the ground or in trees. Terrestrial locomotion consists of a series of short, ungainly hops, but intertree movement is a graceful glide that starts near the top of a tree and ends on another trunk less than 1 m (3 ft) above the ground. Although glides up to 48 m (160 ft) occur, an average "flight" covers 20 m (66 ft) from tree to tree.

This species relishes the typical squirrel diet of nuts, acorns, and conifer seeds supplemented by fruits, buds, lichens, fungi, and sap. Insects
and an occasional bird egg round out the menu. Like other squirrels, it
may hoard food for winter use, but such behavior is not well documented.

Copulation takes place in early spring, and 2–4 young are born 37–42
days later in May. The hairless newborn weigh 5–6 g (0.2 oz) and are
poorly developed; eyes and ears are closed, and the toes are still fused
together. Although the patagium is obvious at birth, the tail is cylindrical
and not flattened, as in the adult. Eyes do not open until day 31, and
youngsters do not leave the nest until day 40. The young squirrel is totally
weaned after two months yet often remains with the mother for another
month or so. A northern flying squirrel generally rears only a single litter
each year.

Most owls and many hawks prey on the northern flying squirrel, and
martens, weasels, coyotes, lynx, and domestic cats commonly eat this
rodent. Occasionally, a flying squirrel dies after its patagium becomes
snagged on barbed wire—a fate shared with some bats. Most live less
than four years in the wild.

Suggested References: Wells-Gosling 1985; Wells-Gosling and Heaney
1984.

Southern Flying Squirrel *Glaucomys volans*

Measurements: Total length: 220–257 mm (8.7–10.1 in); tail length:
85–115 mm (3.3–4.5 in); hindfoot length: 26–33 mm (1.0–1.3 in); ear
height: 13–18 mm (0.5–0.7 in); weight: 50–75 g (1.8–2.6 oz).

Description: The large eyes, flattened tail, and loose fold of skin from
ankle to wrist identify this rodent as a flying squirrel. The silky coat
appears brownish along the back and creamy white below. Belly hairs
of the southern flying squirrel are completely white, unlike those of the
northern flying squirrel, which are dark at the base. Next to the least
chipmunk, the southern flying squirrel is the smallest squirrel in the Great
Lakes region.

Southern flying squirrel

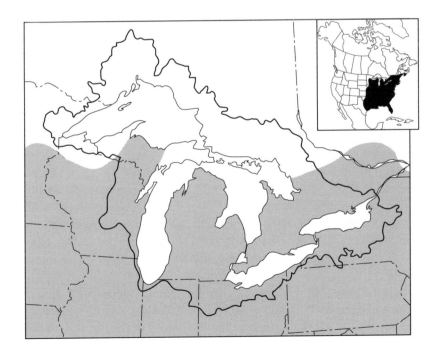

Natural History: This species lives in eastern North America, from Florida to the northern Great Lakes. It prefers open, deciduous woodlands that contain a few shrubby thickets scattered among mature trees. The openness apparently makes gliding less hazardous, and the thickets provide cover when the squirrel is on the ground. Along the northern edge of its range, it occupies mixed coniferous/deciduous woods, especially when deciduous trees dominate. Extensive forested tracts, small woodlots, and even urban parks offer potential home sites.

The southern flying squirrel uses tree cavities, particularly abandoned woodpecker holes, as its favored nest site and occasionally evicts songbirds that were already nesting in the hollow. Entrances are only 40–50 mm (1.6–2.0 in) in diameter, just large enough for this tiny squirrel but too small for any fox or gray squirrel that is searching for a new home. The flying squirrel carefully prepares the cavity with shredded bark and often adds some moss, leaves, or feathers. Although it occasionally builds an external nest of sticks and leaves, it uses such exposed nests to a lesser extent than other tree-dwelling squirrels. Each southern flying squirrel has a primary nest that acts as a daytime retreat and several secondary nests that offer temporary cover while feeding or defecating.

This species is strictly nocturnal and active throughout the night. Although it occasionally uses torpor during winter, it does not store large amounts of fat and does not hibernate. As many as 20 squirrels huddle in a single nest in winter, but they generally lead solitary lives in summer.

The southern flying squirrel does not truly fly, of course; instead, it glides downward from one tree to another. Typical horizontal glide distances are only 6–9 m (20–30 ft), but the squirrels are capable of covering 30 m (100 ft) or more. Although its large eyes are well suited for seeing in dim light, this squirrel can make mistakes. For example, I often capture these beautiful creatures in mist nets that are set to catch bats; I typically place my nets over streams that are 12–18 m (40–60 ft) wide, and the squirrels are caught as they glide from one side of the river to the other. Furthermore, I once saw a southern flying squirrel glide 25 m (80 ft) across a road and clang into a rust-colored metal light-pole. With nothing to grab onto, the squirrel slid to the ground, and after a moment of uncertainty, it scampered toward the nearest tree 5 m (15 ft) away.

Nuts, especially hickory nuts, are the primary food. Decreasing daylight in autumn triggers food-hoarding behavior, and a flying squirrel methodically stores its treasures in tree cavities or the forks of branches.

Fruit, bark, buds, twigs, mushrooms, and lichens are also eaten, particularly in spring when nut caches are depleted. The southern flying squirrel eagerly feasts on animal matter, such as moths, beetles, nestling birds, bird eggs, and carrion.

Young are born in April or May and again in August after a gestation of 40 days. An average litter contains 3–4 offspring; the neonates are pink, hairless, and blind and weigh 3–5 g (0.1–0.2 oz) each. The gliding membrane is obvious at birth, but the tail does not become flattened until 5 weeks later. Youngsters are weaned after 6–8 weeks and often stay with their mother for another 4–8 weeks.

Southern flying squirrels have lived up to 13 years in captivity and 10 years in the wild, but the average life span probably is closer to 5 years. Owls and house cats are major predators.

Suggested References: Dolan and Carter 1977; Fridell and Litvaitis 1991.

Pocket Gophers
Family Geomyidae

Pocket gophers are medium-sized burrowing rodents, weighing 50–900 g (2–32 oz). They have stout bodies with short legs and little evidence of an external neck. The front feet have five toes equipped with huge claws used in digging; the claw on the middle digit is the longest, and its length may exceed that of the rest of the foot. Although pocket gophers also use the incisors to help loosen hard-packed soil, their lips close behind these curved teeth and prevent dirt from entering the mouth. Small eyes and short external ears are additional adaptations to their subterranean existence.

The common name of the family comes from the existence of two cheek pouches or "pockets" that are used to transport food. Unlike the internal cheek pouches of chipmunks, the pouches of pocket gophers are external structures that open on either side of the face, not into the mouth. These pouches are fur lined and can be turned inside out for cleaning.

The skull of a pocket gopher is flat across the top, and the zygomatic arches are widely flared (fig. 9). The skull is similar in size to that of some squirrels or a small muskrat, but a pocket gopher lacks the pointed post-orbital process characteristic of squirrels and has more teeth than a muskrat. Although beaver and porcupine have the same number of teeth as a pocket gopher, they also have much larger skulls. All pocket gophers have at least one groove running down the face of each incisor, and those living in the Great Lakes region have two; no other rodent in our area has two grooves on its incisor.

Pocket gophers are widely distributed across central and western North America, from Canada through Central America. Today, there are 35 different species in 5 genera. Only 1 species, however, is present in the Great Lakes basin.

Plains Pocket Gopher *Geomys bursarius*

Measurements: Total length: 235–310 mm (9.3–12.2 in); tail length: 63–90 mm (2.5–3.5 in); hindfoot length: 30–37 mm (1.2–1.5 in); ear height: 6–9 mm (0.2–0.4 in); weight: 200–400 g (7–14 oz).

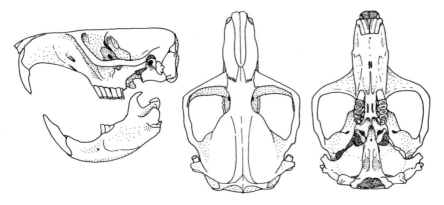

Fig. 9. Skull of a plains pocket gopher, family Geomyidae

Description: The back is usually dark brown to black or gray and the underparts are slightly lighter. The tops of the feet are covered with whitish hairs, and the short, tapered tail is essentially naked. Heavily clawed forefeet, small eyes, short, naked ears, and external cheek pouches readily separate the plains pocket gopher from other mammals in the Great Lakes region.

Natural History: This pocket gopher occurs all across middle North America, from the Gulf of Mexico to the Canadian border, but only enters the Great Lakes basin along its western boundary. Like all pocket gophers, this species spends most of its life underground, and its distribution, not surprisingly, is affected by soil type. It favors a moist, sandy loam and avoids soils high in clay or gravel, as well as those that are very dry or poorly drained. This rodent stays away from dense forests, preferring instead sparsely wooded tracts and open areas. Although alfalfa and clover fields are particularly attractive, the plains pocket gopher generally shuns deeply plowed fields. Average density is 10–12 gophers/ha (4–5/acre).

This species constructs a system of tunnels up to 150 m (500 ft) in length and 30–90 cm (1–3 ft) belowground. Each tunnel is gopher sized, about 10 cm (4 in) across, and has several short side passages, or laterals, that approach the surface; active laterals are used for feeding and old ones for dumping newly excavated soil. In addition, the floor plan generally includes several food storage chambers and a nest area deep in the

Plains pocket gopher. (Photo from the Roger W. Barbour Collection, Morehead State University.)

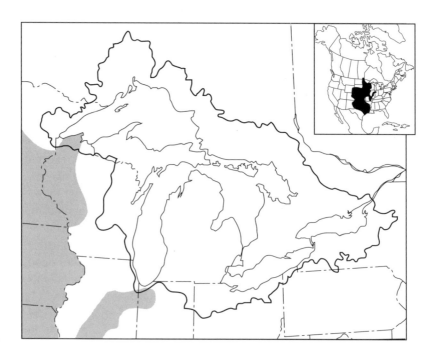

burrow. A nest is a simple hollow ball of dry grass about the size of a large cantaloupe.

When excavating a burrow, the animal loosens hard-packed soil with its massive claws, pulls the dirt under its belly, and then kicks the pile backward with its rear feet. After a sizeable amount accumulates, the gopher turns around and pushes the load down the tunnel, using its head and front feet as a bulldozer. Ultimately it brings the dirt to a surface entrance where it heaps the soil into a conspicuous crescent-shaped mound up to 30 cm (1 ft) high and 60 cm (2 ft) across. Each burrow system is marked by numerous mounds, and the gopher produces 1–3 new ones every day. This species is active year-round and throughout the day, although mound building occurs mostly at sunrise and sunset. The plains pocket gopher is solitary and highly aggressive toward others of its kind.

This species is a strict herbivore feeding primarily on the roots, rhizomes, and bulbs of plants encountered in the rodent's daily digging; stems and leaves are eaten to a lesser extent. Favorite food plants include clover, alfalfa, grass, dandelion, plantain, mullein, and dock. If the pocket gopher encounters a rich food patch, it slices excess items into a convenient size and shoves the pieces into its cheek pouches for transport to a nearby food cache. The pouches are large enough to house more than 50 corn kernels and may accommodate plant stems up to 7 cm (3 in) in length.

Females are monestrous, producing only a single annual litter. The gestation period is unknown, but one female gave birth 51 days after capture. Most births occur in April or May, and litter size averages 3, although it typically varies from 2 to 6. Newborn weigh about 5 g (0.2 oz) and have a total length of 40 mm (1.6 in). Eyes open after 3 weeks, weaning occurs at 5 weeks, and young disperse from the maternal burrow 2–3 months after birth.

The most serious predators are snakes, skunks, badgers, and weasels— all are capable of either entering a gopher's burrow or digging the animal out of its underground home. Although owls are reported predators, the subterranean existence of the plains pocket gopher makes avian predation uncommon. One of these pocket gophers lived more than 7 years in captivity.

Suggested References: Thorne and Andersen 1990; Sudman, Burns, and Choate 1986.

Beavers
Family Castoridae

This is a very small family containing only a single living genus and two species. *Castor canadensis* is the familiar beaver of North America, whereas *Castor fiber* is its Eurasian counterpart. They are the largest rodents in the temperate zone and, worldwide, are second only to the South American capybara in size. Beavers are semiaquatic and have many modifications for life in the water. These include webbed hindfeet for powerful swimming strokes, a flattened tail for use as a rudder, and valvular ears and nostrils that close to prevent water entry. The lips meet behind the large incisors and allow the beaver to gnaw while submerged without simultaneously taking in a mouthful of water. The thick underfur limits heat loss to cold water, and by coating the outer hairs with oily secretions, a beaver prevents cold water from penetrating to the skin. The skull is massive (fig. 10), larger than that of any other rodent in the Great Lakes region, and is about 12–14 cm (5 in) long. The face of each incisor is deep orange, and the ear opening is at the end of an upward-curving, bony tube.

American Beaver *Castor canadensis*

Measurements: Total length: 900–1,200 mm (35–47 in); tail length: 300–400 mm (12–16 in); hindfoot length: 170–195 mm (6.7–7.7 in); ear height: 30–35 mm (1.2–1.4 in); weight: 12–27 kg (26–60 lb).

Description: A large body, proportionately small eyes and ears, and naked, horizontally flattened tail make it easy to identify this rodent. The lush fur is rich brown, both above and below. The webbed hindfeet are much larger than the more dexterous forefeet. The claw on the second toe of the hindfoot is split and used to groom the fur and apply waterpoofing oils.

Natural History: Historically, the beaver occupied much of North America. However, extensive land clearing and fur trapping extirpated many populations and severely depleted others by the late 1800s. Widespread regulation of the fur trade in the early 1900s, along with planned reintro-

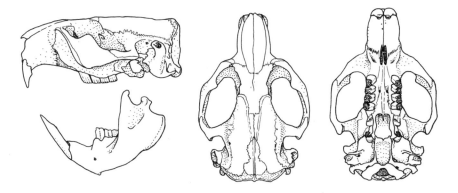

Fig. 10. Skull of an American beaver, family Castoridae

ductions, allowed many populations to recover. Today the beaver is common in the northern Great Lakes basin, but less so in southern areas that are heavily urbanized or extensively farmed.

A beaver prefers slow-moving streams or lakes bordered by young forests containing aspen, willow, or alder. Some beavers live in a burrow hollowed out of the bank adjacent to a river or pond. Most often, however, this rodent builds a lodge by piling logs, sticks, and mud on an island or in a shallow pool. When completed, lodges typically are 1.5–3.0 m (4–10 ft) high and 4–9 m (12–30 ft) in diameter. The inner chamber generally is just above the water line, has two or more underwater entrances, and is cushioned with shredded bark and grass. In addition to providing considerable protection against predators, a lodge also creates a favorable microclimate; air temperature within an occupied beaver lodge remains above freezing even when the outside temperature falls as low as -40°C (-40°F).

If water is too shallow, a beaver engineers a dam across a stream to permanently flood an area and make it more hospitable. Dams up to 30 m (100 ft) in length are known, but most are less ambitious. Although this rodent generally hauls construction materials overland, it sometimes excavates a canal so that larger logs can be floated to the building site. A beaver uses its massive incisors to fell the trees used in lodge and dam construction and relies on its heavily clawed forefeet for burrow and canal excavation.

A beaver is a diurnal herbivore. Main food items are the bark, leaves, and twigs of woody plants, particularly aspen, cottonwood, and willow.

American beaver

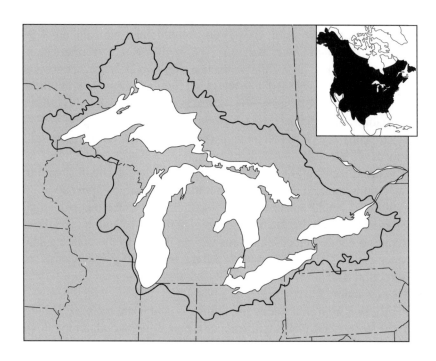

The beaver stores much of its winter food underwater so that, once the pond freezes, the animal still has ready access to fresh bark and twigs. It supplements that tough diet in the summer with a variety of aquatic and semiaquatic plants, especially the roots, rhizomes, and runners of water lilies. In addition, this burly rodent obtains some of its needed nutrients through coprophagy.

Beavers practice monogamy—a rare mating pattern among mammals. Copulation generally takes place underwater, during winter, and the young beavers (kits) are born 107 days later in May or June. Typical litter size is 2–5. The well-furred kits weigh about 450 g (16 oz) at birth and already have their eyes open. Lactation lasts 2–3 months, and a youngster weighs almost 7 kg (15 lb) by six months of age. Kits remain with their parents until their second winter, when they become sexually mature.

Although large size and a secure lodge protect the beaver from most potential enemies, a few, particularly the young, fall to predators every year. The gray wolf and coyote probably are the most consistent predators, but red fox, black bear, wolverine, lynx, and even eagles sometimes attack these rodents. Construction accidents also occur; for example, a beaver occasionally fails to predict correctly the direction of a falling tree and ends up pinned beneath it. Some beavers survive 21 years in the wild, but most do not live beyond 10 years.

Suggested References: Dyck and MacArthur 1993; Jenkins and Busher 1979; Svendsen 1989.

Rats, Mice, Lemmings, and Voles
Family Muridae

This is the largest family of mammals, containing almost 300 different genera and over 1,300 living species—more than one-quarter of all mammalian species belong to this single family. It includes familiar pets, such as the gerbil and hamster, common "field mice," the pestiferous Norway rat, and a host of other rat- and mouselike rodents. Murids are native to all habitable continents, even Australia, and have a long fossil record, dating back 34 million years to the early Oligocene epoch. Older works divide these rodents into a New World family (Cricetidae) and an Old World family (Muridae); however, structural differences between the two are minor, and mammalogists currently emphasize the similarities by grouping them together.

It is difficult to generalize about the life-styles of such a large group of mammals. They are mostly terrestrial or semifossorial in habit, but some species are arboreal or semiaquatic; a few are molelike, live totally underground, and lack functional eyes. In terms of diet, murids are primarily herbivorous and granivorous, yet most take animal flesh to a certain degree. A few are largely carnivorous; for example, the grasshopper mouse of western North America specializes on insects, and the water mouse of Central America subsists on fish and aquatic invertebrates. In general, murids are nocturnal and active year-round. Most weigh less than 200 g (7 oz), although a few exceed 1 kg (2.2 lb); the largest is the slender-tailed cloud rat of the Philippines that occasionally reaches 2 kg (4.4 lb).

In the Great Lakes region, there are 13 native species of murids. In addition, the house mouse, Norway rat, and black rat are three commensal species that followed European settlers to North America. The black rat probably was common in the early villages and towns that bordered the Great Lakes, but it is virtually extinct today. The Norway rat, which came to North America in 1775, is larger, more aggressive, and competitively superior to the black rat; consequently, the Norway rat replaced the black rat throughout much of Canada and the northern United States by early in the nineteenth century. Today, black rats occasionally wander off ocean-going ships that visit the Great Lakes, but there is no established population.

Among native murids, there are two major groupings—the Sigmodontinae (formerly part of the Cricetinae) and the Arvicolinae (formerly

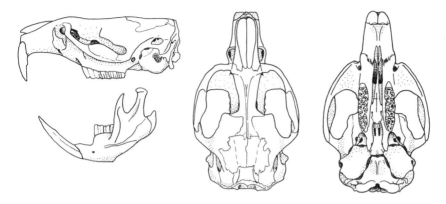

Fig. 11. Skull of a muskrat, family Muridae (Arvicolinae)

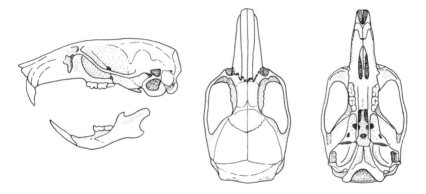

Fig. 12. Skull of a white-footed mouse, family Muridae (Sigmodontinae)

the Microtinae). The Arvicolinae are the voles, bog lemmings, and musk-rat. Although the large muskrat is an exception, members of this group typically have small, beady eyes, short ears partly obscured by fur, and a tail that is less than one-third the length of the head and body. The Sigmodontinae, in contrast, are much more mouselike, and include the deer mouse, white-footed mouse, and harvest mouse; these species have larger, more appealing eyes, taller ears, and a proportionately longer tail than voles and bog lemmings. The teeth also differ. The Arvicolinae have complex ridges of enamel that loop around darker islands of dentine (fig. 11), whereas teeth of the mouselike murids, including the introduced species, generally have two or three rows of small cusps (fig. 12; see also fig. 35 in the Key to Skulls).

One can separate the skull of a murid from that of most other rodents in the Great Lakes basin simply on the number of teeth. There are only three cheek teeth on each side of the upper and lower jaws, plus the incisors, for a total of 16 (see figs. 11 and 12). Only the woodland jumping mouse (family Dipodidae) has such a small number of teeth. These two families, however, differ in the shape of the infraorbital foramen, which is an opening through the bone that forms the anterior margin of the eye orbit. In a murid, this opening generally is narrow at the bottom and broader at the top, i.e., somewhat V-shaped in outline; in a jumping mouse, the opening is an oval whose long axis points down and out.

Western Harvest Mouse *Reithrodontomys megalotis*

Measurements: Total length: 120–152 mm (4.7–6.0 in); tail length: 51–70 mm (2.0–2.8 in); hindfoot length: 14–17 mm (0.6–0.7 in); ear height: 10–12 mm (0.4–0.5 in); weight: 8–15 g (0.3–0.5 oz).

Description: The western harvest mouse is a small rodent with a long tail that is slightly shorter than the body. The animal is blackish brown to brownish gray on the back, but the sides are slightly lighter and often have a hint of orange. Belly hairs are dark gray at the base and tipped with white. The tail is dark gray above and abruptly changes to whitish below. Although this rodent is similar to a small house mouse, the house mouse does not have a distinctly bicolored tail and is usually yellowish brown underneath the body. Also, the western harvest mouse has a longitudinal groove on the incisor that is not present in the introduced house mouse.

Natural History: Although the western harvest mouse is a common resident of central and western North America, it is a newcomer to the Great Lakes region, having reached the southwest corner of the basin within the last 35 years. This small rodent lives in early successional habitats that are grassy to weedy in nature with some tall herbaceous cover and perhaps a few shrubs. Look for the harvest mouse in fencerows, railroad rights-of-way, unused pastures, idle fields, and possibly shrubby edges of streams or marshes.

Western harvest mouse. (Photo by George C. Rinker.)

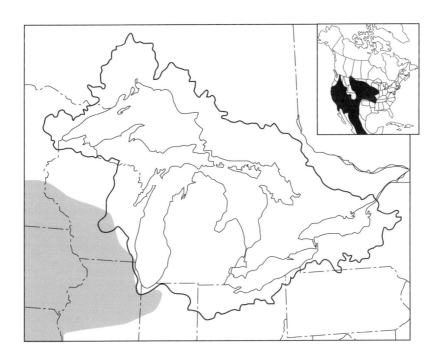

The harvest mouse builds a spherical nest, 8–13 cm (3–5 in) in diameter, from shredded vegetation and hides it in a clump of grass, under a shrub, or beneath a log. The internal chamber is just 3 cm (1.3 in) across, has one or more small openings near the base, and is lined with down from thistle or milkweed. When moving about, a harvest mouse follows grassy runways built by voles and occasionally ventures into low shrubs. This species searches for food primarily between sunset and sunrise and is particularly active on rainy, moonless nights. It is fairly sedentary, and biologists rarely recapture a harvest mouse more than 100 m (330 ft) from where it was originally taken. Home range is about 3,000–5,000 m^2 (0.7–1.2 acres).

This rodent feeds heavily on seeds from grasses, especially foxtail and rye, and it also consumes the sprouts of newly germinated seeds. However, the harvest mouse is omnivorous, and animal matter contributes up to half the diet. Insects are the most common prey, particularly caterpillars, grubs, fly larvae, crickets, bugs, grasshoppers, and leafhoppers, although this rodent also takes earthworms, centipedes, and spiders in very small amounts. During late summer and autumn, it busily collects seeds and caches them for winter use.

The mating season extends from March into October, with a possible lull in activity during midsummer. After a gestation of 23–24 days, the mother gives birth to a litter of blind, deaf, and hairless young. Most litters contain four tiny offspring, each weighing less than a penny, about 1.0–1.5 g (0.04–0.05 oz). Crawling begins by day 5, eyes and ears open by day 12, and the young stop consuming milk by day 24. Those born in the spring reproduce four months later in autumn. In captivity, a female may breed continuously, give birth 14 times in a single year, and produce up to 58 young; however, in the wild, a female probably produces 2–4 litters at most during her lifetime.

As one might expect, predation on these small mammals is quite heavy, and many fall to "mousers," such as the coyote, house cat, red fox, and various weasels. Although nocturnal habits protect the harvest mouse from day-flying hawks, it makes them accessible to night-hunting owls. Life is short and fast in this species, with few surviving for even one year in nature.

Suggested Reference: Webster and Jones 1982.

White-footed Mouse *Peromyscus leucopus*

Measurements: Total length: 145–195 mm (5.7–7.7 in); tail length: 65–95 mm (2.6–3.7 in); hindfoot length: 19–23 mm (0.7–0.9 in); ear height: 15–18 mm (0.6–0.7 in); weight: 16–30 g (0.6–1.1 oz).

Description: The white-footed mouse has large, blackish ears with a narrow white border and bright black eyes. Dorsal fur is russet colored, with a dark band along the midline from head to rump; ventral fur is snow white. The tail is dark brown on top, gradually changes to white or gray on the bottom, and generally lacks a tuft of hairs at the tip. A deer mouse tail, in contrast, shows a more abrupt transition in color and a conspicuous tail tuft. In addition, hairs in the throat area of a white-footed mouse are completely white, whereas such hairs in a deer mouse usually have a gray base. A juvenile white-footed mouse is grayish above for its first 40 days and might be mistaken for a house mouse, but the white underparts of a white-footed mouse are obviously different from the buff-colored belly of a house mouse.

Natural History: The white-footed mouse is broadly distributed through-out central and eastern North America and reaches the northern boundary of its distribution in the Great Lakes basin. It is one of the most common residents of deciduous woodlands, especially where herbaceous cover is moderate and rocks and logs are abundant. Late in the year, subadults may disperse into grassy areas or cultivated fields, but adults avoid such open areas. Shrubby roadsides and fencerows are marginally acceptable habitat.

This mouse builds a nest from just about any soft material that is available—shredded bark, milkweed silk, grass, fur, feathers, etc. The animal gathers the nesting material into a pile, worms its way inside, and creates a cavity by twisting, turning, and pushing. Nests are 15–25 cm (6–10 in) in diameter and have a single opening. The mouse hides the nest under a log, inside a stump, or within a tree cavity, and in rural areas, this species is not above using buildings, including my house, as a sheltered nest site.

A white-footed mouse forages in trees, on the ground, inside mole tunnels, or through passages beneath deep snow. This mouse is generally nocturnal, but daylight activity may occur on extremely cold winter days. Throughout most of the year, this rodent is solitary, and females, in

White-footed mouse

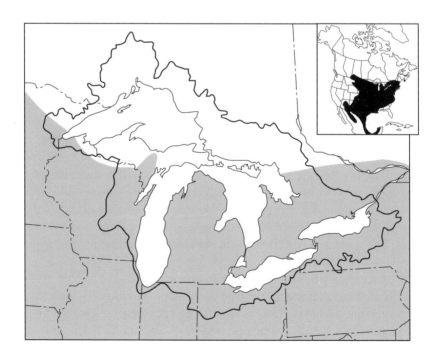

particular, are highly aggressive toward one another. During the winter, they are more tolerant, and groups of 2–6 animals often conserve heat by huddling in a single nest. Some mice save additional energy by becoming torpid for short periods, especially when food is scarce and temperatures are low; however, this species does not store fat in autumn, and it is not a true hibernator.

The white-footed mouse is largely omnivorous. Most of the diet is nuts or seeds from maple, oak, beech, hickory, and pine trees, as well as grasses and cultivated grains. It also eats cherries, grapes, and berries in season. Animal matter, especially grubs and caterpillars, provides needed protein and makes up 30% of the diet. In autumn, a white-footed mouse hoards food for winter consumption; it crams nuts, seeds, and cherry pits into internal cheek pouches and caches these items under logs or in tree hollows.

Mating activity commences in late February or March and, despite a midsummer slowdown, continues into October. Gestation lasts only 22–23 days, and average litter size is 4, varying from 2 to 7. The naked and blind young weigh just 2 g (0.07 oz) at birth. They first leave the nest at 16 days, are weaned by 3–4 weeks, and produce their own litter after 7–10 weeks. All females mate shortly after giving birth, and many produce four or more litters each year. When a nursing mother becomes pregnant with a second litter, the gestation period lengthens to 37 days.

A high reproductive rate counters the extremely high mortality rate that white-footed mice experience. In nature, 98% die in less than one year, although a few crafty individuals live up to two years. Predators include most wild and domestic carnivores, large shrews, snakes, and most species of owl. Squirrels or other mice sometimes steal winter food caches, and such thievery may ultimately result in the mouse dying from starvation.

Suggested References: Lackey, Huckaby, and Ormiston 1985; Schug, Vessey, and Korytko 1991.

Deer Mouse   *Peromyscus maniculatus*

Measurements: *P. m. bairdii*—Total length: 120–165 mm (4.7–6.5 in); tail length: 42–69 mm (1.6–2.7 in); hindfoot length: 15–19 mm (0.6–

Deer mouse

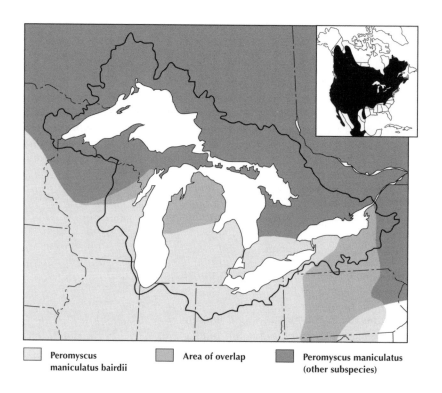

Peromyscus
maniculatus bairdii

Area of overlap

Peromyscus maniculatus
(other subspecies)

0.7 in); ear height: 12–15 mm (0.5–0.6 in); weight: 10–24 g (0.4–0.9 oz). Other subspecies—Total length: 160–205 mm (6.3–8.1 in); tail length: 70–110 mm (2.8–4.3 in); hindfoot length: 18–23 mm (0.7–0.9 in); ear height: 18–21 mm (0.7–0.8 in); weight: 12–24 g (0.4–0.9 oz).

Description: A deer mouse is very similar to a white-footed mouse. Both have bulging black eyes and dusky colored ears rimmed in white. Both are brownish above, with a blackish band running down the dorsal midline, and both have white underparts. However, the back of a deer mouse generally is darker, and the middorsal band is less well defined. The tail of each species is bicolored, brown above and white below, but the colors change abruptly in a deer mouse compared to the gradual blending seen in a white-footed mouse. Also a deer mouse has a more distinct tail tuft than does its white-footed cousin.

There are four subspecies of deer mouse in the Great Lakes region; one of these, the prairie deer mouse (*P. m. bairdii*), is the only subspecies found south of a line running from central Wisconsin, across Michigan to southern Ontario, and on to New York along the southern shore of Lake Ontario (see map). In these southern areas, the prairie deer mouse typically has a smaller ear, hindfoot, and tail than does a white-footed mouse. In other subspecies of deer mouse, the ear is generally *larger* than that of a white-footed mouse, and the tail in these subspecies averages 50% of the deer mouse's total length, compared to only 42% in a white-footed mouse.

Natural History: This small rodent has one of the most extensive ranges of any North American mammal and occupies a diverse assortment of habitats across the continent. In the Great Lakes basin, the prairie deer mouse (*P. m. bairdii*) dwells in open areas, such as meadows, cultivated fields, pastures, and sand dunes along the lakes. Our other subspecies, in contrast, favor forested habitats, but they also venture into shrubby areas, regenerating clear-cuts, and recent burns. In addition, a student of mine trapped these animals amid the rubble of an abandoned limestone quarry in northern Michigan. Here, the animals lived within tiny forest "islands" separated by sterile bedrock; each island was simply a clump of birch saplings growing out of a jumbled pile of quarry debris.

Each deer mouse rests during the day in a baseball-sized nest made from grass and leaves and lined with bits of fur, feathers, or shredded plant material. In grasslands, this rodent generally hides its nest in a small burrow, but woodland dwellers may also nest in a rotting stump, under a log, in a tree hollow, or snug inside a cabin. A deer mouse lives alone during the warmer months, but small groups of 2–5 often share winter quarters. This small mammal forages above or below the snow and is active throughout the cold season, although it occasionally uses daily torpor. Those living in wooded habitats are active climbers, spending as much time in trees as on the ground.

A deer mouse forages mostly at night and relies heavily on olfaction for locating food. This species is a true omnivore, with animal flesh accounting for 15–55% of the diet even in winter. Although it sometimes eats earthworms, insects are preferred, especially crickets, springtails, beetles, moth larvae, and grubs. Plant food is largely seeds gleaned from foxtail, crabgrass, ragweed, sorrel, and other herbs. In agricultural areas, corn, soybeans, and wheat are common foods, whereas berries, wild cherries, nuts, and conifer seeds are important in forests. Fresh food is sparse in winter, and the mouse supplements its meager find with seeds cached during the autumn.

Most deer mice become sexually active in March, but those in northern regions often delay breeding until April or May. The start of mating activity coincides with rising spring temperatures, renewed insect activity, and fresh plant growth. Spring births occur 22–23 days after copulation, and litters typically contain four altricial young weighing 1.5 g (0.05 oz) each. Mothers mate soon after giving birth, and gestation of this second litter is about 35 days. Consequently, at the end of five weeks, she drives off grown young from the first litter just in time to suckle the next. An adult raises as many as four litters each year, if she lives long enough, whereas a mouse born in spring produces only 1 or 2 litters during her first summer.

Even though a captive individual may live eight years, most deer mice do not survive longer than 10–12 months. They are a staple in the diet of snakes, owls, and many mammalian predators, including weasels, foxes, wolves, domestic dogs, and house cats. Many are trapped or poisoned by farmers protecting crops and homeowners unwilling to share nest sites.

Suggested References: King 1968; Klein 1960.

Allegheny Woodrat *Neotoma magister*

Measurements: Total length: 360–440 mm (14–17 in); tail length: 156–200 mm (6.1–7.9 in); hindfoot length: 39–45 mm (1.5–1.8 in); ear height: 25–28 mm (1.0–1.1 in); weight: 260–450 g (9–16 oz).

Description: The head and back of a woodrat is grayish mixed with some black, but adults often display a cinnamon-orange color late in summer. The sides are lighter than the back, and the belly and feet are white. The tail is furred and distinctly bicolored, dark above and white below. This species is the same size as a Norway rat, but a woodrat has softer fur, larger ears, more prominent eyes, and a less pointed snout. Moreover, the scaly tail of a Norway rat is not bicolored, and its belly is grayish brown, not white.

Natural History: This species dwells only in the eastern United States and closely approaches our region in northwestern Pennsylvania, where the Allegheny Mountains abut the Great Lakes basin. Woodrats favor rocky outcrops complete with caves, narrow crevices, and talus slopes. Occupied caves generally are dry and have numerous cracks and fissures in which the rodent hides. The surrounding vegetation in woodrat country is usually deciduous forest.

Unlike other woodrat species, the Allegheny woodrat does not build an enclosed house of twigs and sticks. Instead, it weaves an open nest of coarse woody fibers obtained from red cedar trees and wild grape vines or from the inner bark of hemlock and basswood trees. A nest is similar in shape to that of a blue jay or grackle, but slightly larger, ranging from 30–70 cm (12–28 in) in outside diameter; the inner depression is padded with grass, feathers, or other soft material. The solitary woodrat generally locates its nest on a rocky ledge or in a tight crevice but occasionally moves into an abandoned building, particularly in winter.

This species is a true "packrat" and accumulates all sorts of useless things in and around its home—bones, feathers, plastic wrap, broken glass, cigarette butts, bottle caps, etc. A woodrat always uses specific toilet areas some distance from its nest, and piles of feces up to 5 cm deep may develop. Foraging activity occurs primarily from dusk to dawn and peaks near midnight. The diet consists of a variety of vegetable matter, including mushrooms, leaves, seeds, and acorns and other nuts, as well as fruits, such as wild grapes, black cherries, and apples.

Allegheny woodrat

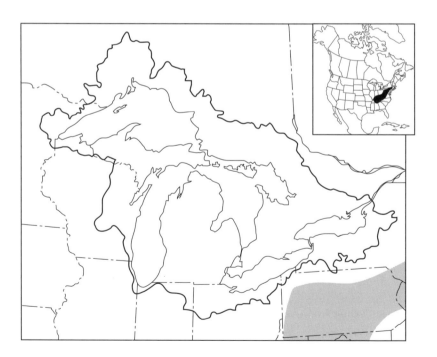

Mating probably begins in very late winter and continues throughout summer. During the breeding season, both sexes have an elongate patch of bare skin along the abdomen that may be 130 mm (5 in) long and up to 7 mm (0.3 in) wide. Strong-smelling secretions exuded by glands in this area often discolor the surrounding fur and presumably act as a sex attractant. After locating each other, the male faces his prospective mate, and both individuals rear back on their hindlegs and proceed to push and scratch with their forefeet. These "boxing" matches apparently are a form of courtship. The bouts occasionally are interrupted by periods of relative inactivity, during which the animals place their forefeet on the shoulders of the other woodrat and excitedly twitch their vibrissae. Copulation eventually occurs, and the female gives birth to a litter of 1–4 pups about 35 days later.

At birth, a pup is naked and helpless, weighs about 15 g (0.5 oz), and is 100 mm (3.9 in) in length, including tail. Fine, silky hair is apparent by the fifth day. Eyes are fully open by 20 days, and the youngsters explore their surroundings shortly thereafter. Young woodrats are almost full-grown after 14 weeks, yet sexual activity does not occur until the spring following birth.

The only confirmed predators of the Allegheny woodrat are the great horned owl and Cooper's hawk, but predation by day-flying hawks is probably not a common threat to night-active woodrats. Skunks, weasels, foxes, and snakes prey on other species of woodrat and presumably attack this rodent as well. Maximum longevity in the wild is about 4 years.

Suggested References: Hayes and Richmond 1993; Poole 1940.

Southern Red-backed Vole   *Clethrionomys gapperi*

Measurements: Total length: 120–160 mm (4.7–6.3 in); tail length: 31–50 mm (1.2–2.0 in); hindfoot length: 17–21 mm (0.7–0.8 in); ear height: 12–16 mm (0.5–0.6 in); weight: 15–35 g (0.5–1.3 oz).

Description: The southern red-backed vole is named for a distinctive, broad, chestnut brown stripe that runs along its back from head to tail. Face and sides appear yellowish brown, whereas ventral hairs are black at the base and tipped with white. In general, voles have much shorter

Southern red-backed vole

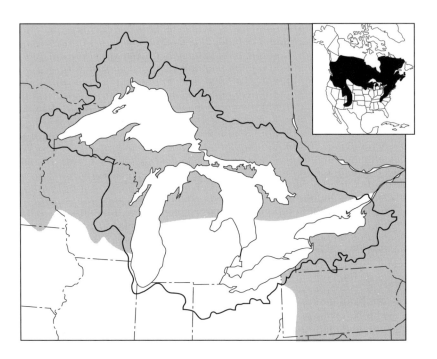

tails than comparably sized mice or rats, and the sparsely furred tail of the southern red-backed vole is only 25–30% of the animal's total length. This vole is somewhat similar to a white-footed or deer mouse, but the short tail should clinch the identification.

Natural History: The southern red-backed vole inhabits the cool, moist forests of boreal North America, from Nova Scotia to British Columbia. Although coniferous forests are preferred habitat, deciduous or mixed coniferous/deciduous woods are acceptable. It frequents lowland habitats, such as swamps of cedar, spruce, or tamarack, and also lives in upland forests, especially if standing water is nearby. The forest floor in this vole's home is a thick carpet of litter with numerous rotting stumps, exposed roots, and moss-covered logs. Typical population density is 4–40/ha (2–16/acre) and is higher in summer than winter.

The red-backed vole is busy throughout the year and at any time of day, although it is most active at night. While foraging, it stays close to fallen logs or rocks and frequently travels through underground passages excavated by large shrews or various moles. This vole is a good climber and occasionally forages or nests in trees. When resting, it occupies a spherical nest inside a tree cavity or under a log; nests are about 7.5–10.0 cm (3–4 in) in diameter and fashioned from grass, twigs, leaves, and moss. This animal is territorial, excluding other red-backed voles from its home and showing aggression toward other species as well. The woodland jumping mouse, for example, actually avoids areas where the red-backed vole is abundant. Home range for this vole is less than 0.5 ha (1.2 acres).

This species is an opportunistic feeder, changing its diet as the seasons progress. It eats leaf petioles and young shoots in spring, adds fruits and berries in summer, and changes to nuts and seeds in autumn. The subterranean fungus *Endogone* is a dietary staple during warm-weather months. Unlike many squirrels and mice, the southern red-backed vole does not commonly cache food for winter use, but it continues to forage for seeds under the snow and supplements this granivorous diet with tree roots and bark when needed. Throughout the year, the red-backed vole obtains added protein by preying on insects, but to a lesser extent than commonly seen in mice.

Mating begins in late winter, often before the snow melts, and females give birth to 4 or 5 young after a brief gestation of 17–19 days. Newborn are able to stand by day 4 and are covered with dark gray hair by day 8;

eyes open by day 15. Body mass is 1.9 g (0.07 oz) at birth and increases to 12 g (0.4 oz) 17 days later at weaning. Youngsters become sexually mature after only three months. Breeding of these promiscuous mammals continues into October, and most adult females manage to rear 2 or 3 litters each year.

A cold snap in late autumn, before an insulating blanket of snow covers the ground, results in much mortality; similarly, in spring, melt-water forces the voles to abandon their subnivean tunnels, exposing the animals to harsh weather and increased mortality. The southern red-backed vole falls to day-hunting hawks as well as night-hunting owls. Weasels, particularly the ermine, are important predators, along with dogs, foxes, coyotes, bobcats, and house cats. Average life span in nature is 10–12 months, with a maximum reported longevity of 20 months.

Suggested Reference: Merritt 1981.

Heather Vole   *Phenacomys intermedius*

Measurements: Total length: 135–155 mm (5.3–6.1 in); tail length: 25–40 mm (1.0–1.6 in); hindfoot length: 17–19 mm (0.7 in); ear height: 12–16 mm (0.5–0.6 in); weight: 30–45 g (1.1–1.6 oz).

Description: The heather vole has the small eyes and short, partially hidden ears that are typical of voles in general. This species is rather plump looking with grayish brown fur across the back and sides and silvery gray feet and underparts. Especially in adults, there often is a touch of yellow or orange on the face and rump. The meadow vole and rock vole are similar; however, the meadow vole lacks the colored nose, and the tail of a rock vole usually is longer than that of a heather vole. Some individuals are difficult to identify on skin characteristics alone, but identification is easy using the dental traits outlined in the account on the rock vole.

Natural History: The heather vole ranges across most of Canada. In the United States, it lives primarily along the western mountains at higher elevations, but it also is found in northeastern Minnesota. Biologists most frequently encounter this rodent in dry, coniferous stands having a dense understory of blueberry, sheep laurel, or other member of the heath

Heather vole. (Photo by J. Bristol Foster.)

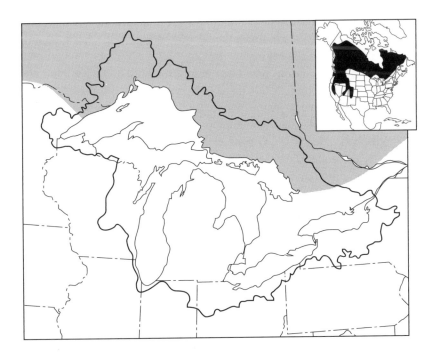

family. Although some heather voles live in shrubby areas bordering forests or in mixed coniferous/deciduous woods, such animals are probably dispersing individuals, particularly subadults, that were forced out of preferred habitat by older and stronger voles.

The heather vole is primarily nocturnal, and it is more active on rainy, foggy, or windy nights than on nights that are clear and calm. Like all voles, this species remains alert throughout winter and continues to forage beneath the snow. A heather vole leads a solitary life for most of the year, but up to five animals conserve energy by sharing a communal nest in the winter. Such a nest is made from twigs and lichens and placed beneath a shrub; the overlying snow hides the nest from predators and provides excellent insulation. In the summer, a nest is generally underground, often beneath a log, stump, or tree root, and is connected to the surface by a short entrance tunnel, less than 1 m (3 ft) in length.

Summer foods are primarily leaves and fruits gleaned from such plants as willow, bearberry, blueberry, soapberry, and dwarf birch. Winter food is mostly bark, along with buds and a few seeds. A heather vole hoards food to a greater extent than other vole species, and it does so at all times of the year. Although a summer food cache consists of leaf and stem cuttings heaped near a burrow entrance, a winter cache is a bunch of twigs piled near a log or the base of a tree. Large winter hoards contain over 1,200 twigs, each measuring 20–60 mm (0.8–2.4 in) in length and 1–6 mm (0.04–0.24 in) in diameter.

The mating season lasts from May to August, and an adult female produces up to three litters each year. Embryos develop for 19–24 days before the mother gives birth to 2–8 offspring. Newborn weigh 2.0–2.7 g (0.07–0.10 oz) each and appear wrinkled and pink. Youngsters right themselves after 6 days, eyes open at 14 days, and weaning begins shortly thereafter. Females born early in the year breed during their first summer, although males wait until the following spring.

Raptors, such as the rough-legged hawk and short-eared owl, are the best documented predators. Starvation, disease, and cold probably result in some deaths over winter. The heather vole avoids most traps devised by mammalogists, making it difficult to obtain information on longevity, but one study suggests that only 6% of the heather voles are still alive after one year.

Suggested References: McAllister and Hoffman 1988; Nagorsen 1987.

Rock Vole or Yellow-nosed Vole   *Microtus chrotorrhinus*

Measurements: Total length: 140–185 mm (5.5–7.3 in); tail length: 42–60 mm (1.7–2.4 in); hindfoot length: 18–22 mm (0.7–0.9 in); ear height: 13–17 mm (0.5–0.7 in); weight: 30–45 g (1.1–1.6 oz).

Description: Small black eyes, short ears, blunt nose, and short tail mark this rodent as a vole. The rock vole appears dark brown to yellowish brown across its back and dull gray along the belly and feet. The most distinctive trait is a yellowish to reddish orange face and nose, and some individuals have a dull yellowish wash on the rump as well. A meadow vole lacks the colored patches, and the heather vole is generally smaller bodied with a shorter tail.

Dental characteristics are often useful in distinguishing among the rock vole and similar species. In all voles, the chewing surface is bordered by a line of enamel that juts in toward the middle of a tooth and back out again following a zigzag pattern (see fig. 11); technically speaking, that part of the border that projects inward, toward the midline of the tooth, is called a reentrant angle. In the heather vole, the inner margin of each *lower* cheek tooth indents more than halfway across the entire tooth, whereas reentrant angles on the outer side are obviously smaller. In meadow and rock voles, the inner angles on the lower teeth are less extensive than in a heather vole and generally similar in size to the outer angles. To distinguish between rock and meadow voles, look at the *upper* cheek teeth; there are four reentrant angles on either side in a rock vole, but a meadow vole has only three (see fig. 35 in the Key to Skulls).

Natural History: The rock vole lives from the western shore of Lake Superior across to Labrador, and down to Tennessee, following the Appalachian Mountains. As its common name implies, the rock vole inhabits talus slopes and rocky outcrops, primarily in forested areas with moderate canopies. It prefers moist sites where mosses or forbs are abundant and generally where open water is nearby, either on the surface or in the subterranean environment. Other small mammals commonly encountered in rock vole habitat are deer mice, red-backed voles, short-tailed shrews, and masked shrews. Throughout its range, the rock vole exists only in small, discontinuous populations.

Biologists know little about its behavior, other than it is most active just after dawn and that much of its activity is restricted to hidden pas-

Rock vole

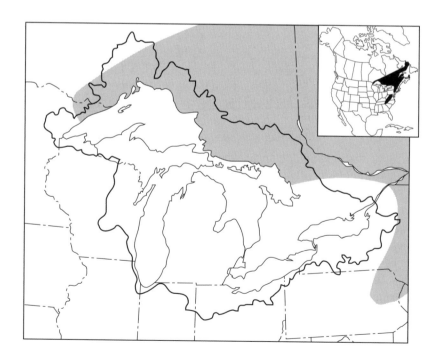

sages beneath the rocks. A rock vole feeds mostly on broad-leaved plants, eating the flowers, fruits, leaves, and even the stems of such species as bunchberry, blueberry, mayflower, goldenrod, and violet. It also consumes seeds, fungi, ferns, and moss in small amounts. Insects—particularly the larvae of beetles, butterflies, and moths—make up about 6% of the diet.

Although litter size varies from one to seven, most females give birth to four young. Older females have larger litters than those giving birth for the first time, and litters in the Great Lakes region commonly are larger than those in more southern populations. The gestation period is probably 19–21 days. Many members of this genus, including the rock vole, undergo a postpartum estrus, and copulation occurs shortly after giving birth; consequently, a female may nurse one litter while gestating another. An adult probably gives birth 2 or 3 times between March and September.

The only documented predators of this uncommon species are the bobcat and a few snakes. Nothing is known about its life span.

Suggested References: Christian and Daniels 1985; Kirkland and Jannett 1982.

Prairie Vole   *Microtus ochrogaster*

Measurements: Total length: 125–155 mm (4.9–6.1 in); tail length: 30–40 mm (1.2–1.6 in); hindfoot length: 18–22 mm (0.7–0.9 in); ear height: 11–14 mm (0.4–0.6 in); weight: 30–60 g (1.1–2.1 oz).

Description: A prairie vole is basically dark brown to black along the head and back. However, dorsal hairs are tipped with either black or brownish yellow, and the mixture of the two gives the animal a salt-and-pepper look. Ventral fur is usually yellowish brown or tan. This species is most similar to the meadow vole; however, a meadow vole has a more evenly colored back and a silvery or grayish belly. In addition, a prairie vole has five small bumps (plantar tubercles) on the hind feet, compared to six in the meadow vole, and a female prairie vole has three pairs of mammary glands, compared to four pairs in the meadow vole. The teeth differ as well; the prairie vole has two reentrant angles on the last upper molar, whereas the meadow vole has three.

Prairie vole. (Photo by John and Gloria Tveten.)

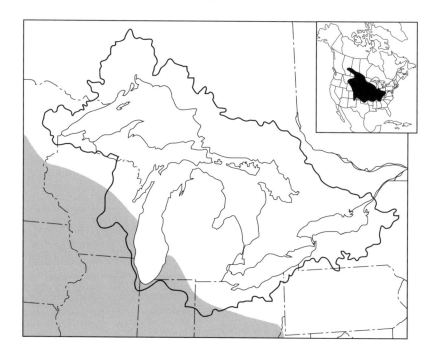

Natural History: This grassland-loving species is a common resident of North American prairies, but it enters our area only in the southwest. It lives in ungrazed pastures, fallow fields, weedy areas, and fencerows. One may find it along highway and railroad rights-of-way, and occasionally in cultivated fields of soybean or alfalfa. In the Great Lakes region, the prairie vole and the meadow vole sometimes occur in the same area; when this happens, the prairie vole occupies drier sites with shorter and more varied vegetation than those that harbor the meadow vole.

A prairie vole is most active in the first few hours before sunrise and in the hour after sunset. It wanders over a small home range, generally less than 0.1 ha (0.25 acres) in extent, following a network of runways constructed through the grass. These "mouse highways" radiate from burrow entrances to feeding areas and allow the vole to move under the protection of a grassy canopy. A burrow is generally short and shallow, and the animal uses it for feeding, caching food, or hiding a nest of dried grass. Especially in summer, a prairie vole often opts for an aboveground nest located beneath an old board or clump of vegetation.

This species is almost totally herbivorous. Even though grasses top the menu, the prairie vole also eats dicotyledonous plants such as alfalfa, dandelion, fleabane, plantain, and clover. It adds a few fruits in summer and autumn, and moss, roots, and bark are eaten in winter and early spring. Seeds contribute to the diet throughout the year.

Unlike most mammals, the prairie vole is monogamous. A male and a female share the same nest and home range, remain together for life, and seldom take a new partner when a spouse dies. The male aggressively excludes other males from the pair's home range and helps care for the young.

Although most breeding takes place from late winter to late autumn, mating occurs year-round if the winter is unusually mild. The long breeding season allows an adult to give birth as many as five times in a single year. Gestation lasts for three weeks, and an average litter contains four offspring. A youngster weighs 2.9 g (0.1 oz) at birth, opens its eyes by day 9, and is weaned by day 17. After 5–6 weeks, a young female mates for the first time, and she typically produces a smaller litter than older, more experienced females.

Many carnivorous animals find prairie voles to be a dependable food source. The Virginia opossum, short-tailed shrew, coyote, red fox, least weasel, and house cat are common mammalian predators, whereas the Cooper's hawk, red-tailed hawk, barred owl, screech owl, and shrike are consistent threats from the air. In nature, average life expectancy is less than 11 weeks, although this species has lived up to 3 years in captivity. The prairie vole experiences cyclical changes in population size every 2–4 years, and densities differ by over 90% between population highs and lows.

Suggested References: Getz et al. 1993; Stalling 1990.

Meadow Vole *Microtus pennsylvanicus*

Measurements: Total length: 130–185 mm (5.1–7.3 in); tail length: 35–60 mm (1.4–2.4 in); hindfoot length: 18–25 mm (0.7–1.0 in); ear height: 11–16 mm (0.4–0.6 in); weight: 35–60 g (1.2–2.1 oz).

Description: The meadow vole is dark brown or black along the back and gray or silver on the ventral surface. Ears are small, rounded, and mostly hidden by the lush fur. The head has a blunt nose, beady eyes, and grades into the body without an obvious neck. The tail is not distinctly bicolored. The meadow vole is similar to the prairie vole, which occurs only in the southwestern part of the Great Lakes drainage, and the rock and heather voles, which occur only in the north and east. A prairie vole, however, has a coarser pelage, five plantar tubercles on the rear feet, and a yellowish brown belly, compared to six tubercles and a grayish belly in the meadow vole. The meadow vole lacks the distinct yellow facial color typical of rock voles, and the meadow vole has a longer tail than the heather vole.

Natural History: This robust rodent ranges as far north as Alaska and as far south as Georgia and New Mexico. It is one of the most common small mammals in our region. The meadow vole prefers moist, grassy fields and also frequents marshes and bogs thick with grasses, sedges, and rushes. Even though dispersing individuals sometimes venture into forested habitats, resident populations seldom exist in such sites.

A meadow vole is active primarily at dawn and dusk, although activity may occur at any time of day or night. Home range size averages less than 0.3 ha (0.7 acre), but it is larger in summer than winter and larger in marshes than meadows. Near its nest, a meadow vole aggressively excludes other voles from a smaller territory of 40 m² (430 ft²) or less.

Meadow vole

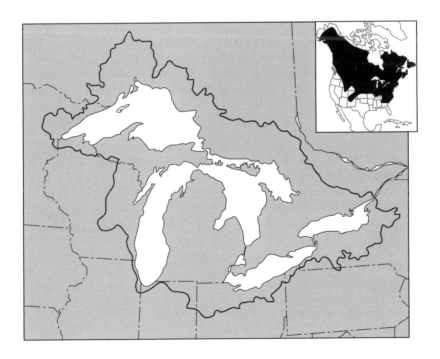

The grassy nest averages 15 cm (6 in) high and twice as wide, usually has a single entrance from below, and is hidden under debris, within a clump of vegetation, or sometimes in an underground tunnel. A meadow vole is a fastidious animal; it is constantly grooming and uses toilet areas located some distance from the nest.

Throughout its home area, a vole carves a complex system of runways through the grass. These canopied trails are originally by-products of the vole's feeding activities, but once formed, the animal constantly maintains them. One often sees remnants of this construction activity in the form of "cuttings"—segments of grass, 2–8 cm (1–3 in) long—piled along the trails. Runways measure 4 cm (1.5 in) across, and interconnect feeding grounds, nest sites, toilet areas, and shallow burrows.

This animal literally lives surrounded by its food, because grasses and sedges dominate the diet. It eats sprouting shoots in spring, goes after tender tips and seeds in summer and autumn, and munches the green basal portions and roots in winter. During warm-weather months, clover, plantain, dandelion, goldenrod, yarrow, and other herbs contribute to the diet. At times of high population density or cold weather, a meadow vole resorts to bark for food and often girdles small trees in the process. Fungi and insects, particularly caterpillars, are eaten in small amounts.

These promiscuous rodents are capable of mating year-round, although reproductive activity often ceases during midwinter. The combination of a short gestation (21 days), postpartum estrus, large litter size (average of six offspring), and ability to breed at any time of year gives the meadow vole an extremely high reproductive potential. In captivity, one female gave birth 17 times in a single year. Even in the wild, an adult female breeds as many as 4–8 times annually. Each altricial newborn weighs 2–3 g (0.07–0.10 oz) and grows rapidly, reaching 14 g (0.5 oz) by 2 weeks of age, when weaning occurs. Youngsters become sexually mature only 4–5 weeks after birth.

Snakes, owls, hawks, shrikes, cranes, and gulls readily eat any vole that they catch. Mammalian predators include the usual carnivores, but also shrews, opossums, and chipmunks. Almost 90% of all meadow voles fail to reach the age of one month, and few, if any, live an entire year. For unknown reasons, a meadow vole population fluctuates dramatically in size, in an approximately four-year cycle.

Suggested References: Madison and McShea 1987; Reich 1981.

Woodland Vole   *Microtus pinetorum*

Measurements: Total length: 110–140 mm (4.3–5.5 in); tail length: 18–24 mm (0.7–0.9 in); hindfoot length: 16–19 mm (0.6–0.7 in); ear height: 10–13 mm (0.4–0.5 in); weight: 20–40 g (0.7–1.4 oz).

Description: The woodland vole has small eyes and ears and a very short tail that is barely longer than the hindfoot. Its compact body is covered with a short, dense pelage that is almost molelike; the back is reddish or chestnut brown, and the belly is mostly grayish. Within the Great Lakes basin, the small body, short tail, and reddish fur easily distinguish the woodland vole from other rodents. Although the southern bog lemming also has a short tail, that species has shaggier fur, a more grizzled coat, and an obvious groove on the upper incisor that is missing in a woodland vole.

Natural History: This rodent inhabits much of the eastern United States, from Texas to Maine, and it ranges across the border into southernmost Ontario and a very small part of Quebec. Unlike the meadow or prairie vole, the woodland vole inhabits a variety of forested areas. This species prefers hardwood forests of oak, maple, and beech, although it is a potential resident of any wooded area—deciduous, coniferous, or mixed. It occasionally lives in orchards and becomes a pest by girdling trees while feeding on bark. This vole is semifossorial, and its burrowing habits require a well-drained, sandy soil overlain by a thick layer of duff.

A woodland vole spends its entire life within a tiny home range, rarely wandering more than 15–30 m (50–100 ft) from the nest. When foraging, it travels either along surface runways, hidden from view by overhanging grass or leafy litter, or through a system of underground tunnels. Each tunnel is about 3 cm (1.2 in) across and located just 5–10 cm (2–4 in) below the surface. The vole generally excavates its own passageways but often incorporates abandoned mole tunnels into the system as well. When resting, a woodland vole occupies a globular nest fashioned from grass and leaves and measuring 15–18 cm (6–7 in) across. This mammal sometimes hides its nest under a log or among tree roots but usually places it at the end of an underground passage. This species is active day or night.

As suggested by its burrowing habits, a woodland vole feeds extensively on roots, tubers, and rhizomes throughout the year. In summer, it

Woodland vole

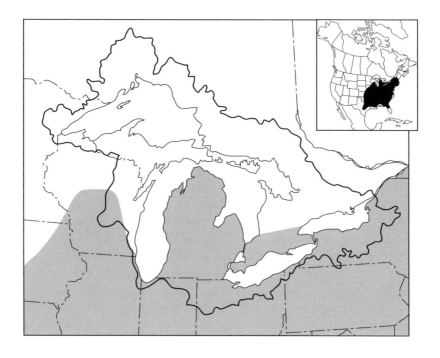

adds clover sprouts, grass stems, fruits, seeds, and nuts. In autumn, it caches tubers and stems in a subterranean chamber for use in winter and often resorts to eating bark before spring arrives. The woodland vole, as well as the meadow vole, practices coprophagy, especially when forced to rely on low-quality foods.

Like the prairie vole, a woodland vole is monogamous; a male and a female share the same nest and home range, and the father helps raise the young. Breeding lasts from January to October, which is plenty of time for an adult female to produce as many as four litters. The young develop *in utero* for 20–24 days and weigh 2–3 g (0.1 oz) at birth. Fine hairs cover the torso by day 5, ears unfold and eyes open by day 8, and the pup walks by day 12, when it weighs 10–11 g (0.4 oz). After 16 days, a youngster begins eating solid food, and it generally is weaned by day 20. A female woodland vole does not conceive until 14–15 weeks of age—much later than either a meadow or prairie vole. The woodland vole has only four mammae, fewer than our other voles, and consequently, most litters contain just 2–4 young.

Average life span is less than three months, and maximum longevity is a little over one year. Numerous hawks and owls prey on woodland voles. Other predators include snakes, foxes, raccoons, weasels, skunks, and opossums. A woodland vole frequently damages trees in nurseries and orchards, and it happily consumes the product of potato farms; consequently, humans have developed a number of poisons to control this rodent.

Suggested References: Anthony, Simpson, and Kelly 1986; Oliveras and Novak 1986; Smolen 1981.

Muskrat *Ondatra zibethicus*

Measurements: Total length: 470–630 mm (18–25 in); tail length: 200–270 mm (8–11 in); hindfoot length: 70–90 mm (2.8–3.5 in); ear height: 20–25 mm (0.8–1.0 in); weight: 0.8–1.5 kg (1.8–3.3 lb).

Description: The muskrat is one of the larger rodents in the Great Lakes region and the largest member of the family Muridae in North America. Dorsal fur is a glossy dark brown, but ventral fur is lighter and often rusty colored. As in many rodents, the tail is naked and scaly, but it also is

Muskrat

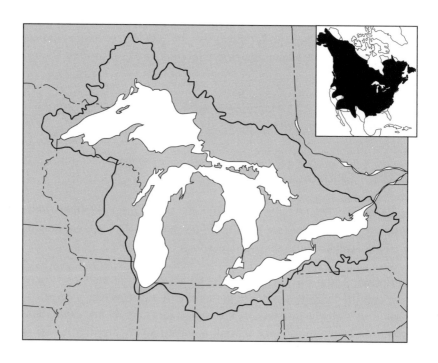

laterally compressed—a condition shared with no other mammal in our area. To aid in swimming, the hindfeet are partly webbed, and the toes have a stiff fringe of hairs. A muskrat is vaguely similar to a beaver, but a beaver is much larger and has a dorsoventrally flattened tail.

Natural History: The muskrat inhabits most of the continent north of Mexico. It is semiaquatic and always lives near slow-moving streams, lakes, ponds, and especially marshes. It prefers standing water that is about 1.5–2.0 m (4–6 ft) deep throughout the year. Water of such depth seldom freezes to the bottom in the winter, but it is shallow enough for rooted plants to thrive.

If the immediate shoreline is elevated, a muskrat uses a bank den. This cozy underground chamber is located just above the waterline and has one or more entrances hidden beneath the water. In marshy areas, however, the rodent constructs a house from mud and emergent plants, especially cattail, bur-reed, and bulrush. A muskrat house has an underwater entrance and is up to 2 m (6 ft) in diameter and 1.2 m (4 ft) in height. It is quite modest compared to a beaver lodge, but the 30 cm-thick (1 ft-thick) walls protect the muskrat from predators, rain, and cold; in winter, average temperature within an occupied house is 20°C (36°F) warmer than the outside air. Such houses seldom last more than a single season, and the muskrat is constantly rebuilding, particularly in spring and early autumn.

This mammal also excavates canals and manufactures "push-ups." Canals are narrow water-filled channels dug through the bottom mud in shallow water; they radiate from the house or den and provide convenient passageways through the marsh vegetation. Early in winter, a muskrat makes a small hole in the ice and pushes up a mass of vegetation gathered from the marsh bottom. Ultimately push-ups freeze solid and provide a concealed resting place and breathing hole that the muskrat keeps open throughout winter.

A muskrat is active at any time of day and feeds primarily on the roots and basal portions of aquatic vegetation, such as cattail, arrowhead, water lily, and various rushes. In addition, it eats small amounts of meat in the form of crayfish, mussels, small fish, turtles, and frogs. A muskrat forages mostly within 15 m (50 ft) of its home and rarely feeds more than 150 m (500 ft) away.

Most young are born between March and October after a gestation of about 25–30 days. The average litter contains 4–8 altricial young that weigh 21 g (0.7 oz) each and measure 100 mm (4 in) in total length. Very young offspring remain in the nest, often covered with dry vegetation, for their first 15 days; at that time, eyes open, and the young muskrat ventures out for its first swim. Although a youngster eats solid food after four weeks, it does not become sexually active until the following spring. The father contributes little care for the young, even though the muskrat generally is monogamous. An adult muskrat produces up to three litters each year.

Humans are the major enemy of the muskrat. Millions are trapped for their fur and meat each year, and many others perish while attempting to cross busy highways. Raccoon and mink are the most important predators, but foxes, domestic dogs, eagles, hawks, and snapping turtles also feed on this large rodent. One young muskrat fell into the window well of my house and could not escape. Maximum longevity in the wild is probably 3–4 years.

Suggested References: Marinelli and Messier 1993; Willner et al. 1980.

Northern Bog Lemming *Synaptomys borealis*

Measurements: Total length: 110–140 mm (4.3–5.5 in); tail length: 18–26 mm (0.7–1.0 in); hindfoot length: 17–21 mm (0.7–0.8 in); ear height: 12–14 mm (0.5–0.6 in); weight: 20–50 g (0.7–1.8 oz).

Description: A northern bog lemming is grayish brown to chestnut brown on the head, back, and sides. It resembles a vole in having small, black eyes, short, rounded ears, and a stubby tail; however, the fur of a bog lemming is obviously longer, and the brownish tail is much shorter than that of any vole within its range. In addition, the bog lemming has a shallow groove running down the face of its upper incisor that is missing in the voles.

Unfortunately, the northern and southern bog lemmings are virtually identical in external appearance and distinguishing between them requires an examination of the *lower* cheek teeth. Just as in voles, the cheek teeth of bog lemmings have zigzag loops of whitish enamel surrounding patches ("triangles") of dentine. The southern bog lemming has a small, isolated triangle of dentine on the outer side of each lower molar, along with three large triangles that form the bulk of the chewing surface. A

From left to right: southern red-backed vole, heather vole, northern bog lemming, and meadow vole. Range map is for the northern bog lemming. (Photo by Robert E. Wrigley, Manitoba Museum of Man and Nature.)

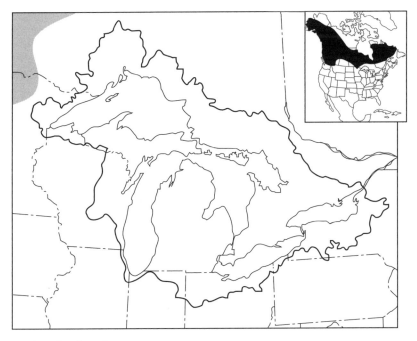

Northern bog lemming

northern bog lemming, in contrast, only has the three large triangles (see fig. 35 in the Key to Skulls).

Females in breeding season may be somewhat easier to differentiate. A northern bog lemming usually has two pair of mammae in the chest (pectoral) region and two more sets between the hindlegs (inguinal). Its southern counterpart retains the four inguinal mammae but only has a single pectoral pair.

Natural History: This truly is a northern species that lives mostly in Canada and enters the United States primarily in the Pacific Northwest and New England. A few specimens, however, have been found in extreme northern Minnesota, not far from the boundary of the Great Lakes drainage. This species apparently lives in sphagnum bogs, conifer swamps, and moist spruce forests, and a few have been captured in wet meadows and in dry, partly wooded areas as well.

It builds and maintains runways through the grass and digs short, shallow burrows. In summer, this rodent fashions a spherical nest from dried grass in an underground tunnel, but in winter the nest usually lays on the surface; snow soon covers it, providing the bog lemming with a favorable microenvironment and shielding this small mammal from its hungry enemies. The northern bog lemming is active throughout the day and all year, never using daily torpor or hibernating. It munches on grasses and sedges and often discards cut stems along its runways. The reproductive season lasts from May to August, and females probably produce two or three litters each year. Typical litter size is four, although the scant data suggest that it ranges from two to eight. Biologists know little else concerning the life of this rare mammal.

Suggested References: Banfield 1974; Clough and Albright 1987.

Southern Bog Lemming   *Synaptomys cooperi*

Measurements: Total length: 110–140 mm (4.3–5.5 in); tail length: 18–24 mm (0.7–0.9 in); hindfoot length: 16–20 mm (0.6–0.8 in); ear height: 10–13 mm (0.4–0.5 in); weight: 20–45 g (0.7–1.6 oz).

Description: This volelike rodent is drab brown along the back and sides and has an overall grizzled appearance and perhaps a hint of red. The

Southern bog lemming

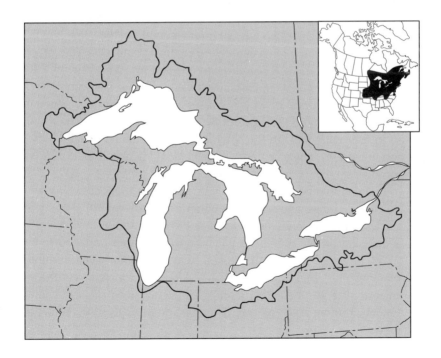

undersides are dark gray interspersed with white. Its most distinctive traits are a tail barely longer than the hindfoot and an upper incisor with a groove on the front lateral surface. These combined features distinguish the southern bog lemming from all other rodents, except its rare cousin, the northern bog lemming. Near the northwestern boundary of the Great Lakes basin, where both species occur, positive identification requires a close look at the lower cheek teeth. Those of a northern bog lemming have only three triangles of dentine, compared to four in a southern bog lemming (see fig. 35 in the Key to Skulls).

Natural History: The Great Lakes basin is the heart of the southern bog lemming's geographic range. This species occupies a variety of habitats, including old fields, clear-cuts, shrubby locations, and upland woods. It also frequents wet, forested sites dominated by spruce, cedar, or tamarack, as well as more open sphagnum bogs. This rodent commonly lives in habitats that seem marginal for the more aggressive meadow vole, and competition between the two may help explain the spotty distribution of the southern bog lemming.

Although occasionally active in daytime, this solitary rodent is mostly nocturnal. It builds runways, similar to those of the meadow vole, through grass, leaf litter, or sphagnum. Piles of bright green fecal pellets and discarded plant clippings, all trimmed to about the same length, are indicative of bog lemming activity. In contrast, a meadow vole produces dark green or brown feces and uneven cuttings. Runways of the bog lemming interconnect feeding sites, toilet areas, and the nest.

Each nest is a ball of woven grasses and sedges with an added bit of fur, feather, or moss. It measures 8–15 cm (3–6 in) in diameter and has up to four entrances. In the summer, a bog lemming usually hides the nest in a grassy tangle or under a convenient log or stump, but its winter home most often is in an underground tunnel.

This docile rodent feeds heavily on succulent monocots. Bromegrass, blue grass, rushes, and sedges are common dietary items that occur along bog lemming runways. When feeding, the animal snips the basal portion of the plant in an attempt to gain the tender upper parts. Surrounding plants, however, prevent the broken stem from falling over, and the bog lemming must make additional snips until the top is in reach. In the process, it leaves behind a number of evenly sized pieces ("cuttings") that are 3–8 cm (1–3 in) long. This species also eats moss, fungi, roots, bark, and such fruits as huckleberry and blueberry. Although primarily herbivorous, bog lemmings occasionally prey on beetles, snails, and other invertebrates.

In the Great Lakes area, breeding generally lasts from March through October, and gestation is 23–26 days. Each neonate has light fur on the head and back; it is blind and helpless and weighs just 4 g (0.14 oz). Ears unfold by day 2, lower incisors break the gum line by day 7, and eyes open by day 11. A pup totally relies on mother's milk for 16 days and is slowly weaned over the next week or so. Shortly after giving birth, a female comes into heat and quickly mates again. Average litter size is three, and an adult female probably produces 2 or 3 litters each year. In captivity, one bog lemming gave birth six times and produced 22 offspring in just six months.

Survival in captivity is up to 29 months, but a wild bog lemming rarely, if ever, lives for an entire year. Like other small nocturnal rodents, a southern bog lemming is common prey for the screech owl, barn owl, and great horned owl. Red fox, gray fox, domestic dog, badger, and house cat are known mammalian predators.

Suggested Reference: Linzey 1983.

House Mouse *Mus musculus*

Measurements: Total length: 150–190 mm (5.9–7.4 in); tail length: 70–95 mm (2.8–3.7 in); hindfoot length: 17–21 mm (0.7–0.8 in); ear height: 11–18 mm (0.4–0.7 in); weight: 15–23 g (0.5–0.8 oz).

Description: A house mouse is grayish or yellowish brown from head to rump; the ventral hairs are most often yellowish brown. The head has a pointed snout, slightly protruding eyes, and large, naked ears. The long, tapered tail has obvious circular rows of scales (annulations) and is very sparsely furred. Although this mouse is somewhat similar to a young deer mouse or white-footed mouse, the buff-colored belly of the house mouse usually identifies it. In the southwestern part of our area, one might confuse a small house mouse with the western harvest mouse, but the harvest mouse has a longitudinal groove on its incisor that the house mouse lacks.

House mouse

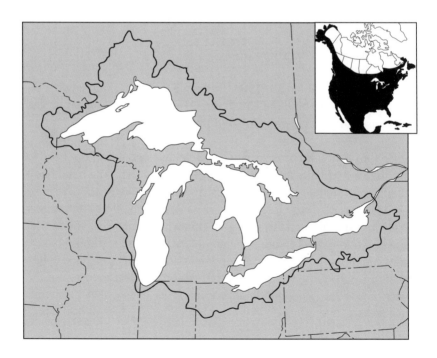

Natural History: The house mouse arrived in North America with the first European explorers and eventually spread throughout most of the continent. It generally lives in close association with humans—in houses, barns, granaries, etc. This troublesome rodent also occupies cultivated fields, fencerows, and even wooded areas, but it seldom strays far from buildings. Some individuals spend the summer in fields and move into barns and houses with the onset of cool autumn weather.

The house mouse lives in small groups containing several females with their young, a dominant male, and possibly subordinate males. The dominant male establishes definite territorial boundaries that all adults in the group defend. For resting and rearing young, outdoor mice usually construct a grassy nest in an underground tunnel. Indoor mice, however, are more creative, hiding their nest under boards or other stored material, inside boxes, cabinets, or walls. Potential nesting material includes discarded paper, mattress stuffing, old rags, and even attic insulation. A house mouse climbs readily, whether it is up a tree or the interior wall of a house. It is active all year, primarily at night, and average home-range size varies from 250 to 1,700 m^2 (0.06–0.40 acre).

This species is truly omnivorous. Most animal matter in the diet comes from carrion and garbage cans, but the house mouse also preys on insects, including cockroaches, caterpillars, and beetle larvae. It eats the seeds of wild grasses, ragweed, and fleabane, and readily consumes corn, wheat, soybeans, and cultivated fruit. An unattended food bowl meant for a pet cat or dog is a virtual cornucopia. A friend once complained that the mice in her apartment ate more dog food each day than her 35-kg (75-pound) pet; the dog, ferocious to most humans, was subordinate to the mice and would not challenge these tiny rodents for access to the food.

A house mouse has a short gestation of 19–21 days and mates again as early as 12–18 hours after giving birth. Even though field mice cease breeding during the winter, those that live in a climate-controlled building probably breed year-round. This species is quite prolific—a female typically produces 5–10 litters each year, and six young per litter is the norm. At birth, each naked youngster weighs only 1 g (0.04 oz). Growth is not particularly rapid; eyes open after two weeks, and weaning occurs a week later. Females first mate when 7–8 weeks old.

Common predators are snakes, hawks, owls, foxes, and coyotes. House cats are particularly adept at catching these wary and speedy mice, and humans understandably inflict tremendous mortality on house mice that pilfer food or reside in buildings. Under captive conditions, maximum reported longevity is about six years, but the average is only two years. Most wild mice succumb after only 12–18 months.

Suggested References: Bronson 1979; Jackson 1982; Mikesic and Drickamer 1992.

Norway Rat   *Rattus norvegicus*

Measurements: Total length: 320–450 mm (13–18 in); tail length: 125–190 mm (4.9–7.5 in); hindfoot length: 30–45 mm (1.2–1.8 in); ear height: 16–20 mm (0.6–0.8 in); weight: 200–490 g (7.1–17.3 oz).

Description: The Norway rat is a medium-sized rodent, slightly larger than a red squirrel but a little smaller than a gray squirrel. The coarse dorsal fur is various shades of dark brown, becoming lighter on the sides, and there may be an inconspicuous dark stripe running down the back. The underside is usually grayish. The ears are large and naked, and the tail is long and scaly with distinct annulations. This rat looks a little like the eastern woodrat, but that species has a white belly and a more heavily furred tail.

Natural History: The Norway rat, like the house mouse, is a commensal rodent introduced to the Great Lakes region during historical times. The species apparently originated in Asia, reached Europe by the mid-1500s, and arrived in North America about 1775. Today, the Norway rat thrives in cities and towns, particularly in areas with poor sanitary conditions; it frequents sewers, garbage-strewn alleys, unkempt buildings, and other locations that provide refuge and a steady food supply. In farming communities, it congregates in barns, silos, grain bins, and the like. Some rats inhabit cultivated fields in summer but migrate to buildings for the winter.

When foraging, the Norway rat generally travels on the ground, but it also is a good swimmer and climber. It is primarily nocturnal and prowls about a home range that is only 50 m (160 ft) in diameter. In rat society, a large male is dominant, maintaining preferential access to food, water, and resting sites. Females actively defend group resources against strangers and often nest together. A nest consists of leaves, paper, rags, and

Norway rat

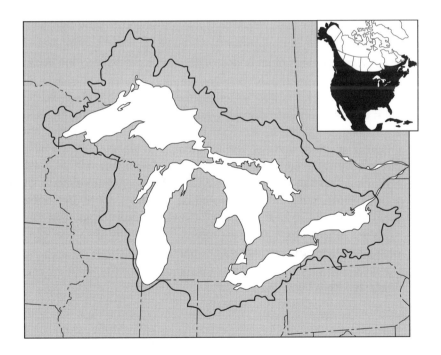

sticks piled on the ground floor of a building or in a short, shallow burrow.

This rat consumes just about anything produced on a farm—grains, fruits, vegetables, eggs, poultry, piglets, lambs, etc. A Norway rat is capable of catching fish and small rodents and readily eats carrion. In cities, this rodent thrives on the uneaten and spoiled food constantly discarded by humans.

The dominant male continuously guards a harem of females and aggressively prevents other males from mating. Each female becomes receptive for a period of about 20 hours, every 4–6 days, and birth occurs just 21 days after copulation. Births are more common in spring and autumn, but this rat breeds at any time of year. A female experiences a postpartum estrus, often mating within 18 hours of giving birth, and easily produces 6–8 litters each year.

Although one prolific female produced 22 offspring in a single birth, most litters contain 8 or 9 young that are blind and hairless. An average newborn weighs 6 g (0.2 oz), but, as in many mammals, birth weight is greater in small litters and lower in large ones. After 15 days, the pup is fully weaned, and it finally leaves the nest at three weeks of age. A young female mates for the first time when 2–3 months old, but her brother must wait until he is older and stronger before challenging the dominant male.

The Norway rat is host for either the vector or causative agent of a variety of deadly human diseases, including bubonic plague. Its penchant for human foods results in large economic losses to farmers, and its gnawing activities damage wiring, pipes, and walls. Consequently, the Norway rat is the target of such persistent control measures as trapping, poisoning, and habitat elimination. Although the aggressive nature of this species protects it from most would-be predators, there are reports of Norway rats as prey for a variety of owls, weasels, dogs, and cats. In "nature," as many as 95% of the rats die before they are a year old, and only a few live into their second or third year.

Suggested Reference: Jackson 1982.

Jumping Mice and Their Allies
Family Dipodidae

The family Dipodidae is an interesting group that contains the jumping mice, birch mice, and jerboas. It has a long fossil history that reaches back to the late Eocene epoch, about 38 million years ago, and today there are 51 surviving species divided among 15 genera. Birch mice range throughout much of temperate Europe and northern Asia, whereas jerboas typically inhabit more arid areas, primarily in northern Africa and central Asia. Most jumping mice live in North America, and some mammalogists place these rodents in a separate family, the Zapodidae.

Dipodids are mostly small- to medium-sized rodents weighing less than 200 g (7 oz). When disturbed, they use saltatorial locomotion, bounding along on elongated, kangaroolike hindfeet. Their long tail, often 50% longer than the head and body, acts as a balancing organ while hopping. Most are true hibernators, storing fat during late summer and early autumn and remaining dormant throughout the winter.

In the Great Lakes region, the skull (fig. 13) of a dipodid is similar to that of a murid rodent; however, dipodids have an oval-shaped infraorbital foramen, in contrast to the irregular, often V-shaped, opening of our other small rodents. A pin-sized foramen lies just under the huge infraorbital opening of a dipodid, but this extra hole is lacking in mice, rats, and voles. Moreover, the zygomatic plate of a jumping mouse is horizontal instead of oblique. The upper incisors of our dipodids have a single longitudinal groove on the anterior surface that is missing in other rodents of the Great Lakes area, except the bog lemmings and western harvest mouse. All dipodids have a total of 16 or 18 teeth.

Woodland Jumping Mouse *Napaeozapus insignis*

Measurements: Total length: 210–250 mm (8.3–9.8 in); tail length: 120–155 mm (4.7–6.1 in); hindfoot length: 28–33 mm (1.1–1.3 in); ear height: 16–18 mm (0.6–0.7 in); weight: 19–32 g (0.7–1.1 oz).

Description: Next to the red bat, the woodland jumping mouse is probably the most handsome mammal in the Great Lakes region. The flanks are a bright yellowish orange streaked with black, and they contrast

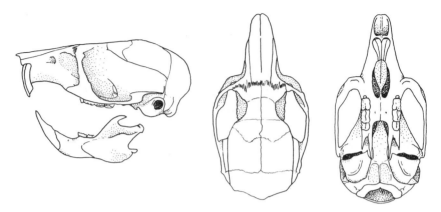

Fig. 13. Skull of a meadow jumping mouse, family Dipodidae

sharply with the broad, dark brown dorsal stripe. The belly hairs are snow white. The tail is longer than the head and body, naked, and scaly; it is distinctly bicolored, dark above and white below, except the most distal part, which is completely white. The hindfeet are about three times longer than the forefeet and seemingly out of place on such a small animal. The huge hindfeet and long tail easily separate this small dipodid from all other rodents in our area except the meadow jumping mouse. However, the meadow jumping mouse is not as brightly colored and lacks the white tail tip. In addition, the meadow jumping mouse has 18 teeth compared to only 16 in the woodland form.

Natural History: The woodland jumping mouse inhabits cool, moist forests in northeastern North America. It favors spruce-fir and hemlock-hardwood associations but also lives in pure deciduous stands. This mammal prefers an environment littered with rocks, logs, and stumps and coated with a lush growth of ferns, grasses, and other plants. Moisture is important, and the jumping mouse is never far from a woodland stream or pond.

This rodent forages primarily at night, but daytime activity occasionally occurs on dreary days. It moves slowly over the forest floor on all fours until startled by a predator. Then, it bounds away on its over-sized hindfeet, in a series of quick leaps, before abruptly halting beneath any available cover. Typical hops are 30–60 cm (1–2 ft) in height and 60–90 cm (2–3 ft) in length. It spends the day in a grapefruit-sized nest of leaves

Woodland jumping mouse

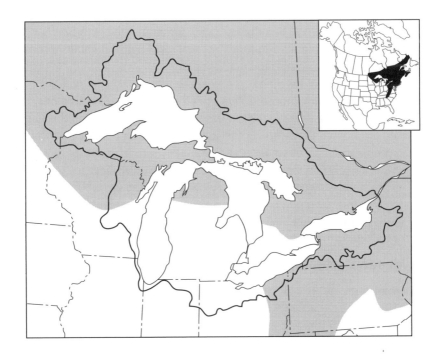

and grass inside a burrow, under a log, or within a brush pile. Home range varies from 0.4 ha to 3.6 ha (1 to 9 acres).

When feeding, the woodland jumping mouse rests on its large hindfeet and manipulates food with dainty forepaws. This rodent relies heavily on seeds throughout the warm-weather months, and it also eats underground fungi and green vegetation, as well as blueberries, blackberries, mayapples, and other fruits. Added protein comes in the form of adult beetles and caterpillars.

In late September or October, this mouse retires to an underground nest, lowers its body temperature, and begins to hibernate. Adults disappear belowground first, followed by the juveniles 1–4 weeks later. When entering hibernation, the mouse lies on its side, places its snout on the belly, brings the hindfeet up alongside the face, and coils the tail twice around the balled-up body. It subsists on stored fat until late April or early May.

In the spring, males leave hibernation first and are ready to mate as soon as the females appear aboveground. Birth occurs 23–29 days after copulation. Although females generally produce five young, litter size ranges from two to seven. The helpless young are naked with loose-fitting, pinkish skin, and the eyes are barely visible as small, dark rings beneath the surface. Young weigh 1 g (0.04 oz) at birth and reach 9 g (0.3 oz) by day 26, when they are well furred and the eyes finally open. Females produce a litter in June and again in August; juveniles probably do not breed until the following spring.

Longevity may be up to 3–4 years, but few data are available. The woodland jumping mouse suffers from the same predators as other small rodents; snakes, hawks, owls, weasels, wolves, and wild and domestic cats are known predators. Some jumping mice do not store enough fat in preparation for hibernation and ultimately starve before spring returns. This is a likely fate for many born in August, because these inexperienced young simply have too little time to prepare for the upcoming winter.

Suggested Reference: Whitaker and Wrigley 1972.

Meadow Jumping Mouse   *Zapus hudsonius*

Measurements: Total length: 180–225 mm (7.1–8.9 in); tail length: 110–140 mm (4.3–5.5 in); hindfoot length: 28–34 mm (1.1–1.3 in); ear height: 12–16 mm (0.5–0.6 in); weight: 12–28 g (0.4–1.0 oz).

Description: Large kangaroolike hindfeet identify this mammal as a jumping mouse. It has a dark brown dorsal band, flecked with yellow, and sides that are yellowish brown, often with a hint of orange. Ventral hairs are white and separated from the darker sides by a narrow yellowish stripe. The tail is longer than the head and body, scaly, and bicolored; its dorsal surface is dark colored all the way to the tip. Although the woodland jumping mouse is similar, it is brighter in color, and the terminal portion of its tail is white.

Natural History: The meadow jumping mouse is broadly distributed throughout the eastern and central United States and ranges across Canada from Labrador to Alaska. It resides in a variety of habitats, including fallow fields, woodland edges, and shrubby thickets. It is most abundant in moist sites containing a lush growth of grasses and forbs, and it particularly favors damp meadows, streamside vegetation, and marsh borders. The number of jumping mice in an area varies considerably from year to year; a typical density is 7–15/ha (2–4/acre), but it reaches as high as 48/ha (12/acre).

The meadow jumping mouse hides its summer nest in a log, under natural debris, or in an underground chamber. A nest is a hollow ball of leaves and grass, about 10 cm (4 in) across, with a single side entrance. This mouse ventures out primarily at night, appearing in daylight only on overcast days. While foraging, it travels slowly through the grass on all four legs or with little hops of just 2–15 cm (1–6 in). When surprised by a predator, it makes several two-footed jumps, up to 30 cm (1 ft) in length, before suddenly becoming motionless or continuing onward with shorter leaps.

Seeds, particularly grass seeds, dominate the diet. It also eats such fruits as raspberries, blueberries, and currants, as well as fungi, beetles, and caterpillars. In spring, when seeds are scarce and fruits have yet to appear, as much as 50% of the diet may come from animal matter.

Starting in late summer, a meadow jumping mouse stores fat for its upcoming winter "sleep." An animal ready for hibernation weighs about 26 g, which is 60–80% heavier than just a few weeks earlier. It spends the winter in a subterranean nest chamber located in well-drained soil below the frost line. Here, it curls into a tight ball and maintains a body temperature of 2–4°C (36–39°F) throughout the winter months. The hibernation period is quite long, lasting from October to late April or early May. Adult males typically enter hibernation first,

Meadow jumping mouse

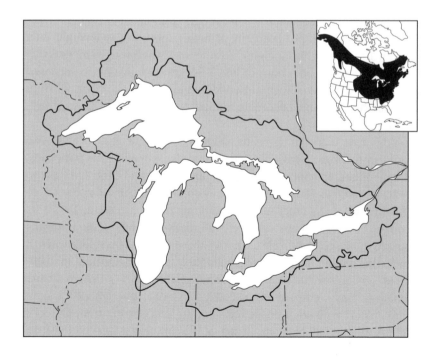

followed by adult females, and then juveniles. In the spring, males emerge before females.

Mating takes place soon after a female comes aboveground, and a litter of 2–8 young is born 18 days later. Individual neonates weigh just 0.8 g (0.03 oz). Although total length averages 34 mm (1.3 in) at birth, the tail is only 9 mm (0.4 in) long. A juvenile becomes fully furred during its third week, eyes open in the fourth week, and weaning occurs shortly thereafter. Adult females usually produce two litters each year.

In the wild, winged predators include the great horned owl, long-eared owl, screech owl, and red-tailed hawk; four-footed enemies include the mink, long-tailed weasel, red fox, and gray fox. Winter is tough on this species, and up to 75% die before spring; many simply run out of fat reserves before fresh food is available. About 9% of the mice that survive their first winter live into their third year. In captivity, maximum longevity is five years.

Suggested References: Hoyle and Boonstra 1986; Whitaker 1972.

New World Porcupines
Family Erethizontidae

Paleontologists have recovered fossil porcupine teeth that date back more than 30 million years to the Oligocene epoch. Despite such a long fossil history, New World porcupines are a small family with only 12 living species in four genera. The family originated in South America and slowly spread northward after the Isthmus of Panama was uncovered during the Pliocene epoch. Today most species still live in Central and South America.

New World porcupines are stocky, slow-moving rodents that could never outrun a determined predator. Instead, the porcupine's coat of sharp spines acts to discourage most animals that attempt to make the rodent into a meal. This antipredator device is not unique to the Erethizontidae; it has evolved a number of times in widely separated mammalian orders, including the egg-laying monotremes (spiny anteaters) and the insectivores (hedgehogs). Among rodents, sharp-pointed hairs also appear in spiny rats (Echimyidae), spiny mice (Muridae), and Old World porcupines (Hystricidae).

New World porcupines spend much of their time in trees and have a number of arboreal adaptations. Feet are long and broad, and the soles are coated with numerous tubercles (friction pads) that provide a slip-resistant surface. The hallux (big toe) is modified into a flexible pad that increases grasping ability. Furthermore, the toes have long, curved claws for digging into tree bark or wrapping around smaller branches. Most South American porcupines, in the genus *Coendou*, even have a prehensile tail.

Only one species lives in the Great Lakes basin, and the large size of its skull separates it from most other rodents, except possibly a few squirrels and the beaver. However, the porcupine skull (fig. 14) lacks the sharp-pointed postorbital processes typical of a squirrel and the elongated auditory tubes found in a beaver. The infraorbital foramen is minuscule in other large rodents, but in the porcupine, this opening acts as a passage for the masseter muscle; consequently, the infraorbital foramen is greatly enlarged and appears as large or larger than the foramen magnum. A porcupine has 20 teeth, 10 above and 10 below.

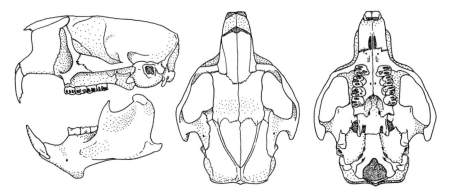

Fig. 14. Skull of a common porcupine, family Erethizontidae

Common Porcupine   *Erethizon dorsatum*

Measurements: Total length: 600–900 mm (24–35 in); tail length: 160–220 mm (6.3–8.7 in); hindfoot length: 75–110 mm (3.0–4.3 in); ear height: 25–40 mm (1.0–1.6 in); weight: 5–14 kg (11–31 lb).

Description: A spiny coat and chunky body make the porcupine easily recognizable. Hollow spines (quills) occur from head to tail, but only on the dorsal surface. Each of these modified hairs is tipped with microscopic barbs pointing toward the base. A single porcupine has 30,000 quills, each measuring up to 75 mm (3 in) long and 2 mm (0.1 in) wide. The basic body color is brown to almost black, although many of the dorsal guard hairs, as well as the spines, have bands of yellow. Next to the beaver, the porcupine is the largest rodent in the Great Lakes region.

Natural History: This spiny mammal inhabits much of North America between the Arctic Ocean and northern Mexico. Although once found throughout the Great Lakes drainage, the porcupine has been extirminated from southern areas now dominated by farmland. In the north, this large rodent lives in deciduous and coniferous woodlands and has a particular liking for stands containing pine and hemlock. A typical density in good habitat is 1–10/km² (2–25/mi²).

In the winter, a porcupine seeks refuge in a cave, decaying log, or hollow tree but does not build a nest. It is not very adventuresome, foraging close to home, following the same trails every day, and continu-

Common porcupine

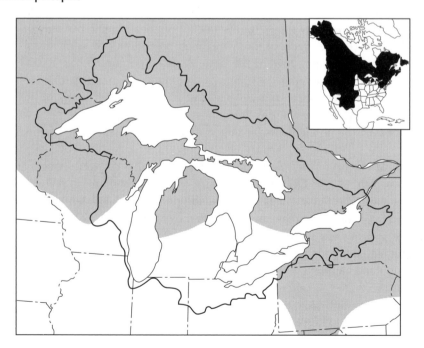

ally using the same den site. Large piles of fecal pellets accumulate near the den, and a distinct odor of stale urine often permeates the surroundings. In the summer, this rodent relies less on a sheltered den and often spends the day resting alone, high in a tree. It is generally solitary and nocturnal.

When approached by a predator, a porcupine faces away from its enemy and raises its spines in a threatening gesture. If the antagonist comes too close, the rodent lashes out with its quill-studded tail. The quills are not thrown, but they easily detach after making contact with a predator's soft flesh. The barbed tips are almost impossible to remove, especially by a four-legged animal, and the quills are pulled gradually inward by incidental muscular movements of the victim. The experience is extremely painful, and death follows if the spines penetrate an artery or vital organ.

The winter diet consists of conifer needles, buds, and the soft inner bark of trees, such as white pine, hemlock, sugar maple, and birch. In the summer, it relies heavily on leaves plucked from basswood, aspen, elm, and birch. In addition to this leafy diet, a porcupine consumes roots, flowers, fruits, seeds, and nuts as they become available.

In November, males fight vigorously over a receptive female, and the combatants often emit high-pitched, catlike screams during the battle. The successful male begins an elaborate courtship that includes grunts and whines and a "dance" while perched on his hindlegs. Unlike most rodents, a porcupine has a lengthy gestation of 210 days and a single, well-developed offspring. A newborn weighs 400–500 g (14–18 oz) and is fully furred. Fortunately for the mother, quills are soft and flexible at birth but harden shortly thereafter. A porcupine is able to see and hear from the moment it is born and walks and climbs in just a short time. It starts eating solid food after two weeks but may remain with its mother for up to six months.

Large body mass and a small litter size usually imply a long life span, and this apparently is true for porcupines. One individual from Michigan lived more than ten years in the wild. Although a number of carnivores occasionally attack this rodent, the fisher has the most success in turning the "porkie" over and attacking its spineless belly. Many of these slow-moving mammals are killed while crossing roads or licking deicing salt from pavement. Some humans consider the common porcupine edible.

Suggested References: Roze 1989; Woods 1973.

Carnivores
Order Carnivora

Mammals belonging to the order Carnivora are primarily flesh eaters, although a number of species, particularly in the bear and raccoon families, have become largely omnivorous. In terms of body mass, carnivores range from the mouse-sized least weasel at 30 g (1 oz) to the gigantic southern elephant seal at 3,700 kg (8,000 lb). Some carnivores, like the seals and sea lions, are highly aquatic and have streamlined bodies and limbs modified into flippers. Most species, however, live on land. Many are capable of climbing trees, and a few, such as the kinkajous, even sport prehensile tails. Carnivores vary from relatively flat-footed (plantigrade), slow-moving animals to swift predators that run on their toes (digitigrade). For example, the cheetah purportedly moves at speeds up to 100 km/hour (60 mi/hour) when chasing its prey.

The teeth of terrestrial carnivores are quite distinctive and designed for a life of meat eating (see fig. 15). The long, conical, recurved canines easily puncture and hold squirming prey, whereas the cheek teeth often are laterally compressed and form efficient cutting blades. In particular, the last upper premolar and first lower molar combine to form a "carnassial pair." When the animal chews, these teeth slide past each other in a shearing action that is reminiscent of a pair of scissors. The carnassial pair is best developed in cats and least developed in omnivorous raccoons and bears.

Paleontologists have unearthed fossil carnivores that date back to the early Paleocene epoch, some 60 million years ago. Today, terrestrial species naturally occur on all continents except Antarctica and Australia, but seals and sea lions routinely visit even these remote land masses.

Worldwide, there are 271 living species classified into 129 genera and 11 families. Eighteen species from 5 different families occur in the Great Lakes basin.

Dogs, Foxes, and Wolves
Family Canidae

This family is familiar to everyone. All canids are essentially doglike, with slender legs, bushy tail, erect ears, and long muzzle. They generally have four toes on the hindfeet and five on the front, although the fifth is reduced and located high on the leg. Canids have blunt claws that cannot be retracted, and they walk and run on their toes—an evolutionary adaptation for speed. In addition, most dogs have a well-developed scent gland that is hidden by a coarse tuft of hairs at the dorsal base of the tail.

The skull is medium size and marked by a long rostrum (fig. 15). Almost all canids have a total of 42 teeth, which is more than most other carnivores. Their strong canines are long and curved, the carnassial pair is well developed, and the posterior molars are built for crushing.

The senses of sight and smell are superb. In the wild, members of this family either stalk their prey and attack with a swift pounce or capture it after a prolonged chase. Small canids hunt either alone or in pairs, whereas larger species more often form packs of up to 30 animals. Although we normally think of wild dogs as attacking other mammals, many canids consume insects, birds, bird eggs, and carrion. One South American species even specializes on crabs.

Canids have virtually a worldwide distribution, and they exist in a variety of habitats ranging from hot African deserts to the frozen Arctic. The largest dog is the imposing gray wolf, which weighs as much as an average man (80 kg [176 lb]). In contrast, the smallest family member is the African fennec fox, which is about the size of a muskrat (1 kg [2.2 lb]). The family dates back to the late Eocene epoch, about 38 million years ago. Biologists currently recognize 34 living species, and 4 of these dwell in the Great Lakes region.

Coyote *Canis latrans*

Measurements: Total length: 1,100–1,300 mm (43–51 in); tail length: 290–390 mm (11–15 in); hindfoot length: 180–220 mm (7.1–8.7 in); ear height: 95–125 mm (3.7–4.9 in); weight: 11–21 kg (24–46 lb).

Description: A coyote reminds one of a small German shepherd dog. The body varies in color, but it usually is yellowish gray with a whitish

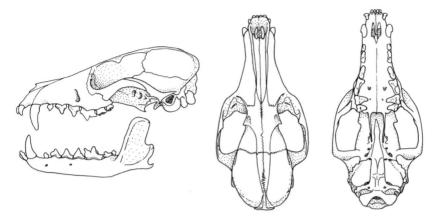

Fig. 15. Skull of a red fox, family Canidae

throat and belly. A dark band extends along the back and onto the tail, which is tipped with black as well. A coyote typically has erect ears and a bottle-shaped tail that is narrow at the proximal end and flares out to its maximum width just before the tip; a domestic dog, in contrast, often has folded ears and generally has a brush-shaped tail of more constant width. When running, a coyote holds its tail below the back, whereas a dog carries its tail straight out or raised. A coyote is significantly larger and more heavily built than any fox but much smaller than a gray wolf. Also, the ears of a coyote are larger in proportion to body size than those of the wolf.

Natural History: The range of this animal now includes all of North America except portions of northernmost Canada. The coyote is most at home in prairies, brushy areas, and wooded edges and least tolerant of unbroken forested tracts. Before European settlement, the coyote was absent from most of the Great Lakes drainage. Today, however, the mosaic of farms and woodlots in the south, as well as the patchwork of forests and clear-cuts in the north, are quite acceptable as coyote habitat. This canid is much more tolerant of human activities than its cousin, the gray wolf, and occasionally is seen near farm buildings and at the edge of towns.

It is most active at night but sporadically ventures out during daylight, especially in the summer. Home range is typically 10–40 km^2

Coyote

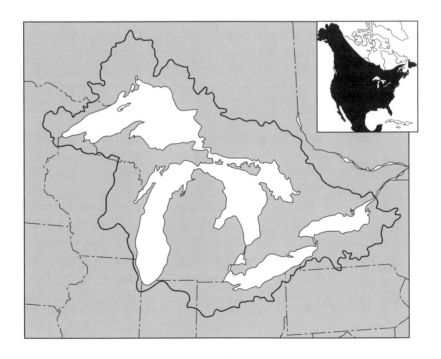

(4–15 mi^2), much larger in males than females, and larger in summer than winter. These monogamous canids hunt as individuals, with their partner, or as a family unit including immature young; they rarely form packs. Individuals slowly stalk mouse-sized mammals until the hunter suddenly pounces, stiff-legged, on the surprised victim. In contrast, members of a pair or family often cooperate when attacking bigger game, such as an old or sick deer. One coyote may drive the deer toward waiting comrades, or they may take turns chasing the animal until it tires. The coyote, like many other canids, is an excellent sprinter and travels at speeds up to 56 km/hour (35 mi/hour) for short periods.

A coyote is opportunistic and feeds on just about any wild mammal. Documented prey include shrews, squirrels, voles, mice, rabbits, hares, muskrats, porcupines, deer, and moose. The coyote rarely kills species larger than itself and prefers, instead, to scavenge off a reasonably fresh carcass. Although 90% of the diet consists of mammals, a coyote occasionally eats birds, snakes, frogs, fish, and even insects. Fruits and vegetables are taken in small amounts, especially in the autumn.

Unlike a domestic dog, a coyote comes into heat only once each year, for 4–5 days, sometime between January and March. Birth occurs in an underground den, 58–65 days after copulation. An average litter contains six pups that are covered by short, woolly hair and weigh about 260 g (9 oz) each. Eyes do not open until day 14. A pup begins eating food regurgitated by the mother during the third week and learns to hunt after it is two months old. Occasionally the father or older siblings from an earlier litter watch over the growing young or bring them food. Even though many young disperse from the natal den after 6–9 months, most do not breed until their second winter. A coyote readily hybridizes with a domestic dog and occasionally with the gray wolf.

Adult coyotes contend with few natural predators, and in our area, only the gray wolf and black bear are capable of dispatching one. Humans are the major threat through hunting, trapping, and poisoning. Although one wild coyote survived almost 15 years in the wild, most do not live longer than 6–8 years. Record longevity in captivity is 18 years.

Suggested References: Bekoff 1977; Schmitz and Kolenosky 1985; Sheldon 1991.

Gray Wolf *Canis lupus*

Measurements: Total length: 1,350–1,700 mm (53–67 in); tail length: 370–475 mm (15–19 in); hindfoot length: 235–295 mm (9.3–11.6 in); ear height: 100–130 mm (3.9–5.1 in); weight: 35–65 kg (77–143 lb).

Description: Dorsal color is most often grayish but may be brown or black; underparts are lighter. The long, bushy tail has a black tip, and the ears are erect. The gray wolf is the largest wild canid, and overall body size alone should distinguish it from the coyote. Compared to a large domestic dog, the wolf has a narrower chest, bigger feet, and longer legs; the fur around the mouth is generally white in a wolf but dark in a dog. In addition, the face of a wolf is framed by a line of long hairs extending downward from the ears, but this fringe is usually missing in a dog.

Natural History: The gray wolf still roams over most of Canada and Alaska. Throughout its range, this species occupies diverse habitats, including tundra, mountains, prairies, and forests. The two prime requisites for wolves appear to be an abundance of large game and minimal interference from humans. Although the gray wolf at one time lived throughout the Great Lakes region, it remains only in the far north of our area, with established populations in Ontario, Minnesota, and Wisconsin. The wolf population in Michigan's Upper Peninsula is small, but steadily growing due to immigration from Wisconsin and Ontario as well as the increased reproductive success of resident animals. Unfortunately, the well-studied population on Isle Royale has declined in recent years and is on the verge of extinction.

Wolves live in packs of 5–9 adult animals, in a rigid social hierarchy led by a dominant (alpha) male. The other males sort themselves out as well, with one individual dominant over the next in rank, who in turn dominates the next, and so on down the line. In addition, females have their own hierarchy, with the mate of the alpha male at the top; all females, however, are subordinate to any male. The pack leader and his mate probably founded the pack some years earlier, and most members are their various-aged offspring. Packs roam over an expansive home range, measuring 50–800 km^2 (20–300 mi^2), and the group often covers 20–50 km/day (12–30 mi/day), mostly at night. Pack members cooperate in defending the territory, raising young, and hunting large prey.

Gray wolf. (Photo by Jim Wvepper.)

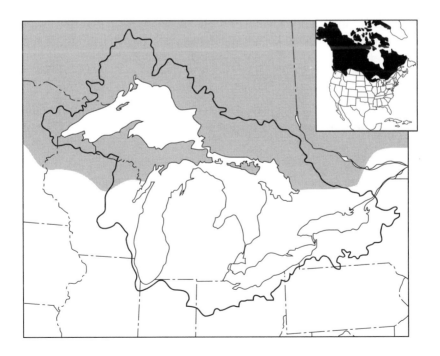

A gray wolf consumes 2–6 kg (4–13 lb) of meat each day, although it is capable of bolting down 9 kg (20 lb) in a single feeding. In the Great Lakes basin, the most likely victims are white-tailed deer and moose. By culling the old, sick, and very young from ungulate herds, the wolf actually stimulates productivity in prey populations, because there is now more food and cover for the healthy, reproductively active, herd members. Up to a third of the diet consists of smaller mammals, such as beavers, rabbits, squirrels, and even mice. The wolf locates its quarry by scent or chance and sometimes by following fresh tracks.

The alpha male and his partner mate for life, and they often are the only pack members to breed. After a 63-day gestation, in late April or May, 5–7 young are born in an underground den. At birth, a pup has closed eyes and floppy ears, but it is well furred and weighs about 500 g (1.1 lb). Eyes open 11–15 days after birth. Pups first play outside the den at 3 weeks, and they are weaned at 6–8 weeks. Pack members bring food to the lactating female and her youngsters or "baby-sit" while the parents hunt. A wolf does not breed until it is 2–3 years old, but it continues to reproduce into its 10th year. On occasion, a gray wolf mates with a large domestic dog or coyote and viable offspring result.

An adult gray wolf fears no predator. Natural mortality in the Great Lakes basin results primarily from starvation, disease, intraspecific strife, and wounds inflicted by uncooperative prey. About 40% of our wolves die at the hands of humans—through trapping, shooting, and chance encounters with automobiles. One gray wolf lived for 16 years in captivity.

Suggested References: Fritts and Mech 1981; Fritts et al. 1992; Mech 1974.

Red Fox   *Vulpes vulpes*

Measurements: Total length: 950–1,050 mm (37–41 in); tail length: 325–400 mm (12–16 in); hindfoot length: 140–175 mm (5.5–6.9 in); ear height: 75–90 mm (3.0–3.5 in); weight: 3.5–7.0 kg (7.7–15.4 lb).

Description: The red fox usually has a handsome red or yellowish red coat with white underparts. The bushy tail is reddish, sprinkled with black, and it becomes white at the very tip. The slender legs and feet

Red fox

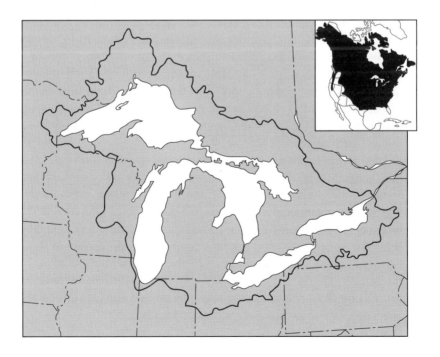

appear clothed in black "stockings" that contrast sharply with the red back. Two color variants, the "cross fox" and the "silver fox" are seen occasionally. The cross fox appears mostly yellowish or grayish brown and has a vague black cross in the shoulder area. The silver fox, in contrast, has a blackish body with a variable amount of white frosting. Both color phases usually retain the black legs and white-tipped tail that are a characteristic of the species.

Natural History: The red fox lives in most of Canada and the United States, but it is scarce in many of the western states. It prefers open country with reliable cover nearby and frequents forest-field edges, brushy fencelines, and the wooded borders of streams or lakes. Although this canid generally avoids dense unbroken forests, it often becomes common after loggers open the canopy. A red fox occasionally lives on the suburban fringe and sometimes occupies large parks, cemeteries, or golf courses.

A male red fox is a loner during much of the year, but he forms a strong pair bond with a single female in early winter and remains with her until the young disperse in late summer. During the reproductive season, life centers around an underground den, which often is a modified woodchuck burrow located in an open hillside or streambank; the den has several entrances, smells strongly of urine, and is marked by discarded food scraps. Outside the breeding season, a red fox simply curls up in a thicket or beneath a brush pile. This canid covers 8 km (5 mi) on average during its nightly search for food and typically occupies a home range of 100–500 ha (250–1,200 acres). A red fox consistently announces its presence to other foxes by repeatedly urinating at various sites throughout its home range.

Even though a mated pair occasionally searches for food side by side, they generally are solitary hunters. The red fox uses its acute hearing and keen sense of smell to locate prey; the predator slowly stalks its victim, attacks with a sudden pounce, grabs the prey with its forepaws, and kills with a quick bite behind the head. Edible small mammals include rabbits, squirrels, mice, and voles, and this fox consistently consumes ground-nesting birds as well. A red fox is somewhat opportunistic, often feeds on carrion, and occasionally takes a snake, crayfish, salamander, and even a cricket or beetle. During late summer and autumn, this carnivore also consumes small amounts of nuts and fruits.

Between early January and mid-March, copulation occurs, and the young develop *in utero* for the next 51–54 days. Although a pup is blind at birth, it is covered by woolly fur and already possesses a white-tipped tail. Most litters contain five pups weighing 100 g (3.5 oz) each. The young begin playing around the den entrance after five weeks and begin their life as meat eaters by eight weeks. A male red fox is a good father; he brings the vixen birds and mice to feed on, keeps a watchful eye over pups when they venture out of the den, and accompanies youngsters on their first hunting trips. By late September, the family begins to disband. Juvenile males leave first, dispersing 30 km (18 mi), on average, away from their parents, but young females typically stay closer to home. Both males and females become sexually mature and mate during their first winter.

The average life span is only 1 year, but a few survive for 5–6 years. Although lynx, bobcat, and coyote sometimes catch a red fox, this agile mammal easily escapes the more heavily built gray wolf. Large hawks and owls threaten only unwary youngsters. Sarcoptic mange is prevalent, often quite severe, and sometimes contributes to the early demise of a fox. In addition, this species occasionally suffers from outbreaks of distemper or rabies.

Suggested Reference: Henry 1986.

Common Gray Fox   *Urocyon cinereoargenteus*

Measurements: Total length: 875–1,050 mm (34–41 in); tail length: 300–390 mm (12–15 in); hindfoot length: 125–145 mm (4.9–5.7 in); ear height: 70–80 mm (2.8–3.1 in); weight: 3.5–6.8 kg (7.7–15.0 lb).

Description: Although slightly smaller on average than a red fox, the gray fox is best distinguished by its coat color. The legs and flanks are cinnamon brown, and the belly is a dirty white. Dorsal guard hairs are black tipped and have a prominent subterminal band of white—a combination that gives the gray fox an overall grizzled appearance. There often is an indistinct blackish stripe going down the dorsal midline and onto the tail. The tail ends with a band of black hairs, not white as in the red fox.

Natural History: The tree-climbing gray fox is a typical resident of deciduous forests throughout North America. However, the Great Lakes

Common gray fox

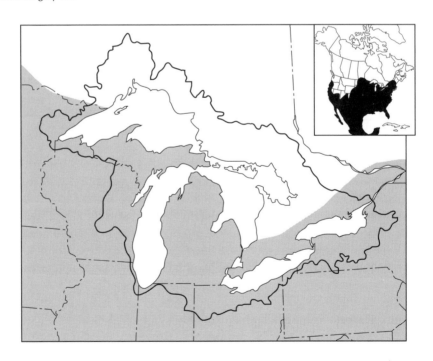

mark the northern edge of its range, and the species is generally uncommon in our area. It lives where woodlands and farm fields are well mixed, in wooded swamps, and in bottomland forests. Unlike its red cousin, the gray fox shuns cultivated fields and open meadows, spending its time in woods and thickets instead.

For shelter, it uses a rocky crevice, brushpile, or underground den. A gray fox rarely digs a den from scratch, preferring instead to modify an abandoned woodchuck or badger burrow. Unlike other canids in our area, the gray fox readily climbs trees to search for prey, rest, or escape predators. Using its highly curved foreclaws, this digitigrade mammal is capable of ascending 18 m (60 ft) up a tree trunk, as well as safely jumping from branch to branch. It is most active at night or near dawn and dusk.

The gray fox feeds primarily on small mammals—cottontails, in particular, with a sprinkling of voles, mice, and squirrels. Birds, grasshoppers, and crayfish are taken in smaller amounts. Plant material often is the largest component of the diet in late summer and autumn when this carnivore consumes large amounts of corn, apples, grapes, and nuts.

The vixen and her partner generally mate in March and remain together until the young are raised. Although the gestation period is not known, it probably is similar to the red fox, or about 53 days; hence, most young likely are born in May. Females typically produce four pups in their single annual litter. Each neonate weighs about 86 g (3 oz), has closed eyes, and is cloaked in a fuzzy fur coat. The father brings food to his mate during lactation and generally helps out with the young. Offspring are weaned by the end of three months and start hunting by themselves after four months. In late summer, youngsters disperse up to 84 km (50 mi) from the natal den, and most become sexually mature during the first winter.

Lynx, bobcat, coyote, and large raptors are likely enemies, but documented predation on this species is virtually nonexistent. As in the red fox, canine distemper or rabies occasionally infect a gray fox, and a localized epidemic may result. Unfortunately, humans are responsible for most mortality with hunters and trappers annually harvesting up to 50% of the population in some areas. Most die before their second birthday, but a few survive 6–10 years in the wild.

Suggested Reference: Fritzell and Haroldson 1982.

Bears
Family Ursidae

The bear family includes only nine living species in six genera. It contains the largest terrestrial members of the order Carnivora, including two species that weigh as much as an automobile. The largest grizzly bears live in southern Alaska and weigh up to 780 kg (1,700 lb). Male polar bears also approach this size and measure close to 3 m (10 ft) in length. In contrast, the lesser panda is the smallest family member at a diminutive 5 kg (11 lb).

Bears are typically large bodied and short legged. Ears are small and rounded, and the tail is practically nonexistent. The shaggy coat is mostly one color, although the giant panda has a striking mixture of black and white. Bears are plantigrade, shuffling along on the soles of their heavily clawed feet. Some, such as the polar bear, are excellent swimmers, and most species have mastered tree climbing. Although the heavily built skull possesses large canines (fig. 16), the carnassial pair is poorly developed and reflects the omnivorous diet of most ursids. Anterior cheek teeth are rudimentary, and one or more often are missing (see the lower jaw shown in fig. 16). Molars are quite large and generally elongate.

Bears live in warm tropical forests, open plains, high mountains, and even the frozen Arctic. They are native to North America, Europe, and Asia and live in the Atlas Mountains of North Africa and the Andes of South America as well. Their fossil history extends back to the late Eocene epoch, about 50 million years ago. Only one species lives in the Great Lakes region.

Black Bear *Ursus americanus*

Measurements: Total length: 1,250–1,800 mm (49–71 in); tail length: 80–125 mm (3.1–4.9 in); hindfoot length: 200–280 mm (7.9–11.0 in); ear height: 100–135 mm (3.9–5.3 in); weight: 110–230 kg (240–500 lb).

Description: The glossy coat is mostly black or dark brown, both above and below. The muzzle is usually a light brown, and often there is a V-shaped, whitish, throat patch. Ears are short, rounded, and erect, and eyes are small relative to the head. Each of the large feet has five toes,

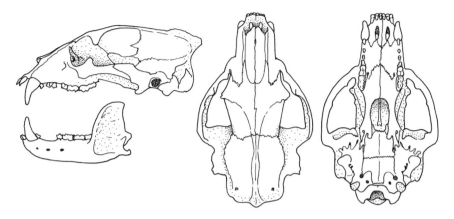

Fig. 16. Skull of a black bear, family Ursidae

and each of these is equipped with a nonretractile claw. The tail is inconspicuous. This bear is the largest carnivore in our region.

Natural History: Before European settlement, the black bear lived in most forested areas of the continent, but its current range is much reduced. A black bear prefers dense coniferous or deciduous woods having a thick understory and lacking continual disturbance from humans. Both upland forests and swampy woods are acceptable.

This bruin is largely crepuscular, even though it may be seen at any time of day. Home range is quite large, but variable, ranging from 500 to 15,000 ha (1,200–37,000 acres). The home ranges of different bears often overlap, especially near a concentrated food source, but individuals usually avoid each other. During warm weather, it rests under a convenient shrub or leafy branch. However, as winter approaches, it excavates a simple den under a stump or brush pile or takes advantage of a ready-made cavity in a hollow tree or cave. Unlike foxes or the coyote, a bear often lines its den with moss, grass, and leaves.

A black bear stores a considerable amount of fat during the autumn in preparation for a unique form of dormancy termed "carnivorean lethargy." From early November until mid-April, it remains in its den and does not feed or drink. During this time, heart rate declines from 40 to 10 beats/minute, and body temperature falls slightly from the usual 38°C

Black bear

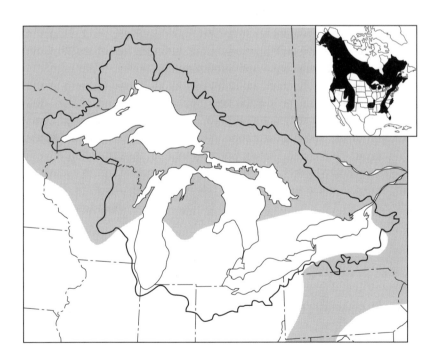

(100°F) to 33°C (91°F). This small drop in temperature allows the inactive bear to survive until spring on fat reserves alone. Unlike true hibernators, such as ground squirrels and bats, the black bear is not helpless during winter torpor and is capable of shuffling away when disturbed.

About one-quarter of the diet is animal food—mostly carrion and invertebrates. The powerful bear rips open rotting logs or overturns rocks searching for ants and grubs and raids the nests of bees and wasps looking for developing young and stored honey. It occasionally takes a fish, rabbit, or mouse, and, even more rarely, a deer fawn or moose calf. A black bear relishes fruit of all kinds, including raspberries, cranberries, strawberries, apples, grapes, and cherries. In late summer and autumn, beechnuts and acorns are dietary staples.

Breeding occurs in early summer, but the embryo does not implant in the uterine wall until the sow begins winter dormancy. Ultimately, cubs are born in January or February, some 210 days after mating. Two is the average litter size. Despite the bulk of the mother, newborn are tiny, weighing about 275 g (10 oz), or roughly the size of a small rat. Neonates are sparsely furred, and the eyes are closed. Claws are apparent but still soft. Even though cubs are weaned in September, they remain with the sow and share her winter den before finally dispersing in their second spring. Females mature sexually at 3–4 years of age, but males generally wait another year. An adult female only mates every other summer.

Most youngsters that die do so from malnutrition, whereas death of an adult is usually human related. If a cub survives its first two years, it has an excellent chance at a long life. Wild bears survive up to 13 years, and some zoo-dwelling individuals have prowled their cages for up to 26 years.

Suggested References: Horner and Powell 1990; Pelton 1982.

Raccoons and Their Allies
Family Procyonidae

Although only one procyonid (the common raccoon) inhabits the Great Lakes basin, there are 18 living species divided among six genera. The family is exclusively American in distribution, occurring from Canada in the north to Argentina in the south. Procyonids are primarily forest dwellers that live from sea level to elevations of 3,000 m (10,000 ft) or more. Ancestors of modern procyonids lived in North America as early as 30 million years ago, during the Oligocene epoch.

In general, they are small- to medium-sized carnivores that usually weigh between 1 and 14 kg (2 and 30 lb). Body color varies slightly from gray to reddish brown and generally is quite uniform on any one animal. Facial markings, however, are common, and the long, furry tail generally has alternating light- and dark-colored bands. All procyonids are good climbers and use the tail as a balancing organ when moving through trees. These mammals walk on the soles of their feet (plantigrade), and each foot has five clawed toes.

The large canines (fig. 17) identify these animals as members of the order Carnivora, even though procyonids are mostly omnivorous. Not surprisingly, therefore, the carnassial pair is weakly developed at best and provides minimal shearing action. As in the Ursidae, anterior cheek teeth are not well developed, but those in the rear are large and best suited for crushing. A proportionately short rostrum, along with six upper and six lower cheek teeth on each side, easily distinguishes this family from other carnivores in the Great Lakes region.

Common Raccoon *Procyon lotor*

Measurements: Total length: 700–925 mm (28–36 in); tail length: 220–260 mm (8.7–10.2 in); hindfoot length: 110–125 mm (4.3–4.9 in); ear height: 50–60 mm (2.0–2.4 in); weight: 6–20 kg (13–44 lb).

Description: The raccoon is a stout-bodied mammal with a distinctive black "mask" across an otherwise white face. Overall the pelage appears brown to black and mixed with a lighter yellowish brown. A raccoon has a cylindrical tail with five or more circular bands of light brown fur

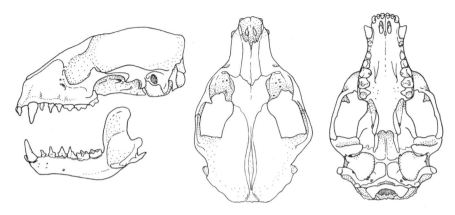

Fig. 17. Skull of a common raccoon, family Procyonidae

alternating with darker rings. The overall body size, striped tail, and black mask are sufficient to identify this species.

Natural History: This carnivore occupies much of Central and North America, from Panama to southern Canada. In the Great Lakes basin, a raccoon lives in or near wooded areas, often near a stream or pond, and is more abundant in hardwood stands than pure coniferous forests. It is an intelligent and curious species that is capable of surviving in remote wilderness, farm woodlots, or urban parks.

Home range is as small as 5 ha (12 acres) in suburban areas and as large as 50–300 ha (120–740 acres) in rural locations. In the summer, activity begins near sunset, peaks before midnight, and continues at a reduced pace until sunrise or even later. Acute hearing and sensitive forepaws make nocturnal hunting much easier. In addition, a raccoon has a highly developed "tapetum lucidum" in the inner eye that aids in night vision. After light originally passes through the eye, the tapetum reflects it through a second time, thus increasing the total amount of light available to each receptor in the retina. In the glare of a flashlight, some reflected light bounces totally out of the raccoon's eye and back to the observer, making it seem as if the eyes "glow in the dark."

A raccoon usually dens alone in a hollow tree but occasionally uses a rock crevice, woodchuck burrow, or abandoned building. Tree dens most often are 2–20 m (6–66 ft) high in a sturdy maple, elm, or oak. In

Common raccoon

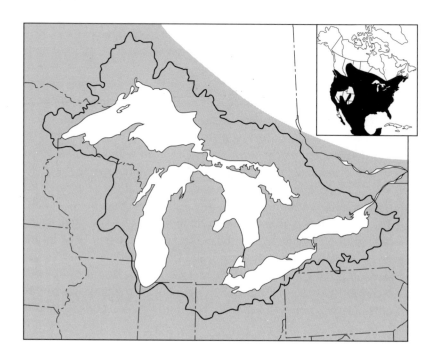

the winter, a raccoon stays in its den for prolonged periods, especially during inclement weather. Even though this mammal stores fat during autumn, it does not hibernate, nor does body temperature fall in a manner comparable to a black bear.

A raccoon is an omnivore par excellence. Although choosy when food is abundant, it eats anything organic if hungry enough. Berries, cherries, grapes, and acorns are commonly eaten and supplemented with corn, soybeans, and oats. This mammal frequently forages along water courses and quickly devours any crayfish that it encounters. Other animal fare includes earthworms, insects, rodents, rabbits, birds, snakes, turtles, frogs, and fish, as well as various eggs. Although a raccoon appears to wash its food, such behavior is not a necessity, because many forage in upland areas away from open water.

Courtship and mating activity begin in early February and continue into March. Fetuses develop for 63 days and are born during the spring warm-up. At birth, a raccoon weighs about 85 g (3 oz), is lightly furred, and already has its identifying mask and tail rings. Eyes do not open until three weeks after birth. The growing raccoon begins taking solid food after nine weeks, yet it continues to suckle for another month or two. Older youngsters accompany their mother on foraging trips, often following her in single file through the underbrush. Young usually den with the mother during their first winter and disperse at the start of the next breeding season, when most become sexually mature. An adult female usually produces a single annual litter containing 2–6 young.

In nature, average life span is 2–3 years, but a few survive up to 13 years; one animal held in a zoo lived more than 22 years. Large owls and carnivores, such as the gray wolf, coyote, red fox, and bobcat are potential predators. Tasty meat and a marketable fur coat mean that many die at the hand of human hunters and trappers. In addition, many raccoons succumb while crossing highways, particularly in the spring when young disperse and males search for mates. Canine distemper and rabies are potential problems, and a major epidemic of raccoon rabies has spread northward from the mid-Atlantic states and has already entered the Great Lakes basin in New York.

Suggested References: Clark et al. 1989; Lotze and Anderson 1979.

Weasels and Their Allies
Family Mustelidae

This family contains the weasels but also many other mammals that, at first glance, are not so weasellike, including the badger, wolverine, and various skunks. Fossil mustelids, about 34 million years old, are known from the Oligocene epoch of Asia, Europe, and North America, and today, mustelids occur on all continents except Australia and Antarctica. These carnivores occupy virtually all terrestrial habitats, and many forage in lakes, rivers, and even the oceans. It is a medium-sized family of 70 recent species divided into 25 genera.

Most mustelids have long, thin bodies and short legs. As in the bears and raccoons, each foot has five toes with nonretractile claws. Toes are fully webbed in the semiaquatic otters, and claws are best developed in the semifossorial badgers. The smallest species, the least weasel, weighs only 30–60 g (1.1–2.1 oz) and is often smaller than the mice and voles that it preys upon; the largest mustelid is the sea otter, and it may weigh as much as 45 kg (100 lb). In most species, strong sexual dimorphism exists, with males typically 10–25% larger than females. Anal scent glands are well developed in most mustelids—extraordinarily so in skunks.

Relative to the braincase, the rostrum is very short (fig. 18). Most species are carnivorous, and the teeth are well suited for a life of meat eating. Canines are long and curved, and the carnassial pair is usually the largest of the cheek teeth.

Many mustelids, especially in temperate regions, utilize "delayed implantation" as part of their reproductive strategy. As in most mammals, fertilization occurs soon after copulation, but the mustelid embryo usually does not undergo continual growth. Instead it ceases development when still a small mass of cells (blastocyst) and does not immediately implant in the uterine wall. The embryo remains in this quiescent state for up to one year, depending on species. Then, in response to changing daylength, the blastocyst resumes normal development, implants, and birth occurs 25–65 days later. At least in some species, this delay allows the young to be born in the spring, when prey are more abundant and environmental conditions are more favorable.

Mustelids are the best represented of all carnivore families in the Great Lakes basin. Today, 10 species from six genera live in our area. An

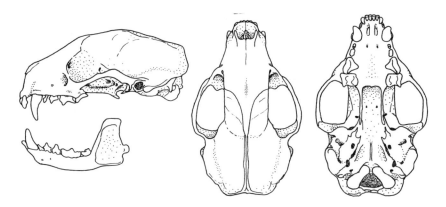

Fig. 18. Skull of a striped skunk, family Mustelidae

additional species, the wolverine, was extirpated from the region in the late nineteenth or early twentieth century.

American Marten *Martes americana*

Measurements: Total length: 550–660 mm (22–26 in); tail length: 170–220 mm (6.7–8.7 in); hindfoot length: 75–95 mm (3.0–3.7 in); ear height: 34–45 mm (1.3–1.8 in); weight: 725–1,500 g (1.6–3.3 lb).

Description: The marten is about the size of a small, slim, domestic cat with long, dense, glossy fur. Dorsal and ventral sides are golden brown to dark brown, except for a light yellowish brown or orange patch covering the throat and chest. Tail, legs, and feet are darker than the back, whereas the face appears lighter, often whitish. The ears are generally rimmed with white. This mustelid is most similar to the mink in size and shape, but the marten is usually lighter in color and has longer fur. Moreover, a mink lacks a white edge to the ears, and a marten has a bushier tail.

Natural History: The American marten lives in forests across much of Canada, but it is absent or uncommon in all but the most remote regions of the United States. Overtrapping and deforestation led to the elimination of the marten from much of the southeastern portion of its original

American marten. (Photo by John and Gloria Tveten.)

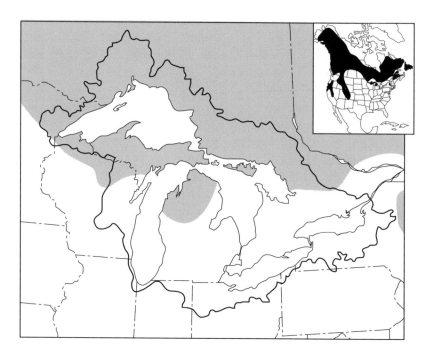

range by the early 1900s. In the Great Lakes region, natural populations of this carnivore thrive in Ontario and Minnesota, and humans have reintroduced the species into Wisconsin and the Upper and Lower Peninsulas of Michigan.

The marten primarily resides in closed, coniferous woodlands underlain by a lush growth of shrubs and forbs and appears less frequently in mixed stands. It occurs in upland forests of spruce, fir, or hemlock, as well as in cedar swamps. Logging clear-cuts, burned areas, and other open sites are useless to a marten.

In pristine habitat, home range averages 1 km^2 (0.4 mi^2) for females and 2–4 km^2 (0.8–1.5 mi^2) for males; in disturbed areas, home range is often 3–4 times larger. This solitary mustelid is active year-round, mainly at night and near dawn and dusk. It moves along the ground on occasion but is an excellent climber and spends much of its time in trees. In winter, it often searches for prey in snow-covered spaces found beneath woody debris. Other than females with young, a marten has no fixed shelter; it rests in a tree hollow, rotting stump, or rock pile or simply on an exposed branch. A maternity den is usually a leaf-lined tree cavity or rock crevice.

Ground-dwelling rodents, especially red-backed voles, are preferred food, although a marten commonly eats red squirrels and northern flying squirrels as well. Other animals that fall prey to the marten include shrews, hares, birds, reptiles, and amphibians. Beetles and bugs form a small, but consistent, part of the diet, and carrion is eaten whenever it is available. In summer, the marten adds fruits, particularly blueberries and raspberries, to its meaty menu.

Mating activity commences in late July or August. Copulation generally occurs on the ground, after a playful courtship lasting up to 15 days. Although fertilization occurs soon after mating, development ceases for 190–250 days. After it resumes, pregnancy lasts only 27 days, and an average litter of 2–3 young is born in late March or April. Neonates are blind and deaf and have a sparse coating of yellow hairs. Even though male and female litter mates weigh about 28 g (1 oz) each at birth, a male is obviously larger after only three weeks of growth. Ears open at 24 days and eyes at 39 days; weaning begins shortly thereafter. After 3–4 months, a young marten reaches adult size but does not attain sexual maturity until 15–24 months of age. Wild females are capable of breeding even in their 12th year.

Record longevity is 17 years in confinement and 13 years in the wild. Predation is rare and comes from owls, eagles, various canids and felids, and perhaps the fisher. A marten is curious and easily trapped, and continued governmental regulation of the harvest is a must if the species is to survive in the Great Lakes basin. Habitat degradation, particularly through extensive clear-cutting, also is a serious long-term threat.

Suggested References: Clark et al. 1987; Sherburne 1993.

Fisher *Martes pennanti*

Measurements: Total length: 800–1,075 mm (31–42 in); tail length: 280–400 mm (11–16 in); hindfoot length: 90–125 mm (3.5–4.9 in); ear height: 40–55 mm (1.6–2.2 in); weight: 2.0–5.5 kg (4.4–12.1 lb).

Description: A fisher is a slender mammal with a pointy nose and bushy tail. Overall, it appears dark brown, but the face, head, and shoulders are washed with gray. The pelage is thick and glossy and somewhat coarser than that of a marten. Ventral fur is dark brown and often has irregular white blotches on the throat, chest, or belly. This species is the largest of the weasellike mustelids in our area.

Natural History: The fisher occurs in a narrow band stretching across Canada and portions of the northern United States. Historically, this species ranged throughout the Great Lakes basin, but unregulated trapping and deforestation led to its extinction, except in remote regions of Ontario and Minnesota. Current populations in Michigan and Wisconsin resulted from successful restocking programs started in the 1950s. Like the marten, a fisher prefers the interior of dense coniferous forests and avoids totally open areas, but it is also capable of living in mature deciduous woods and even young, second-growth forests.

A fisher hunts alone over 15–35 km^2 (6–14 mi^2) of forest. The larger home range of a male extensively overlaps that of several females, but not that of another male. This mustelid is generally diurnal, except during summer when nighttime activity is common. Temporary shelters include hollow logs, rock piles, and abandoned beaver lodges, but more permanent maternity dens usually are found in tree cavities. Although capable of climbing trees, a fisher primarily moves along the ground.

Prey include snowshoe hares, red-backed voles, red squirrels, and various mice and shrews. In addition, the fisher is the major predator of

Fisher. (Photo from the Roger W. Barbour Collection, Morehead State University.)

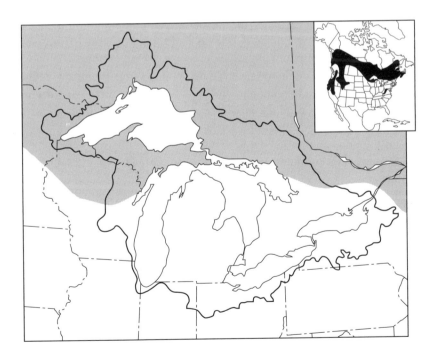

the common porcupine. This carnivore attacks the "porkie" with repeated bites to the face—the only exposed part of that animal not protected by quills. After 30 minutes of such harassment, the spiny rodent weakens from exhaustion, shock, and blood loss and can no longer defend itself. Following the kill, a fisher takes 2–3 days to eat a single porcupine, devouring most of the animal except the quill-covered skin, a few large bones, and the intestines. This mustelid occasionally preys on blue jays and ruffed grouse and readily eats carrion. Despite its name, a fisher does not prey on fish, and its common name may be a corruption of Dutch or French words referring to the polecat, a European weasel. Fruits, such as black cherries and blueberries, form up to 20% of the summer diet.

A female gives birth in March or April, experiences a postpartum estrus, and mates again. The fertilized egg divides until it is a mass of cells about 1 mm (0.04 in) across and then suspends development for 10–11 months. Late the following winter, development resumes, and birth occurs 30 days later; thus, total time spent *in utero* is about one year. An average litter contains three altricial neonates that are sparsely furred and weigh about 40 g (1.4 oz) each. Eyes do not open for 53 days. Youngsters eat meat brought by the parent after two months, effectively kill prey after four months, and disperse after five months at the earliest. A female mates when she is one year old, but her brother waits for his second birthday before breeding.

The size and agility of an adult fisher discourage predation, although large raptors, red fox, bobcat, and lynx probably snatch a few immature fishers. Humans pose the greatest threat to the species through fur-trapping and habitat destruction. A fisher lives as long as 10 years in the wild.

Suggested References: Arthur and Krohn 1991; Powell 1981 and 1982.

Ermine   *Mustela erminea*

Measurements: Total length: 235–330 mm (9.3–13.0 in); tail length: 50–95 mm (2.0–3.7 in); hindfoot length: 28–45 mm (1.2–1.8 in); ear height: 15–22 mm (0.6–0.9 in); weight: 50–155 g (1.8–5.5 oz).

Description: In the summer, the head, back, and outer legs are a uniform chocolate brown, whereas the underparts, including the chin and inner

Ermine in winter pelage

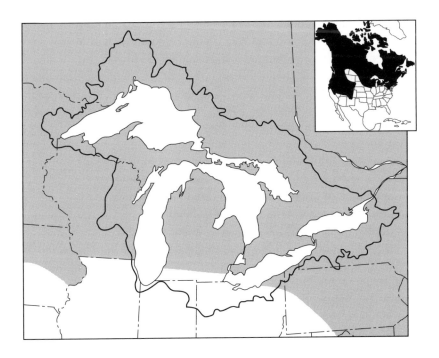

parts of the hind legs, are white. The tail is brown close to the body, but the outer third is black. In the winter, the entire animal becomes white, except for the black tail tip. The ermine is generally larger than the least weasel and smaller than the long-tailed weasel; however, the range of body sizes overlaps that of both species. In contrast to the ermine, the least weasel generally lacks a distinct black end to the tail. The long-tailed weasel's tail is greater than 44% of the head and body (not total) length, whereas the relative length of an ermine's tail is less.

Natural History: This ferocious carnivore lives in the vast boreal regions of Asia, Europe, and North America. In the Great Lakes region, the ermine occupies a variety of habitats, including open forests, riparian woodlands, and shrubby fencerows. It is more common to the north.

Home-range size varies considerably with prey density, but it typically is 21 ha (63 acres) for males and 12 ha (29 acres) for females. Like other weasels, an ermine patrols its territorial boundaries and marks them with secretions from its large anal glands. It defends its home area against members of the same sex, although territories of males and females overlap. While patrolling its homeland, an ermine sequentially visits several nest sites located near favored hunting spots and takes 10–15 days to make a complete circuit of its territory. Each nest is lined with grass and leaves and with fur plucked from unfortunate prey. A typical nest is found under a rock, in a hollow log, or in a rodent burrow.

An ermine hunts mainly at night but forages during the day as well; in winter, it continues to search for prey under the snow and does not hibernate. Mice, voles, and shrews make up 75% of the diet, and most of the remainder comes from lagomorphs, squirrels, birds, and other vertebrates. When hunting, an ermine zigzags through the environment, poking its head under forest debris and into underground tunnels. Each prey is approached slowly at first and then grabbed after a quick rush. The weasel wraps its long body around the victim and dispatches it with a swift bite to the back of the neck or head.

In this species, copulation lasts 2–20 minutes and, in some cases, may be repeated up to five times in one hour. Ermine are born in April or May after an average gestation of 280 days that includes an 8–9 month period of developmental delay. As in other weasels, the longer days of spring probably trigger the resumption of development. A typical litter contains 4–8 naked young weighing about 2 g (0.07 oz) each. The eyes are tightly shut at birth and do not open for 4–6 weeks. A young female mates when only 2–3 months old, often before it is weaned, and well before it reaches adult size at six months of age. In contrast, a male does not gain adult dimensions or breed until his second summer.

Although extremely aggressive when cornered, the ermine falls prey to larger carnivores, such as red fox, coyote, marten, and fisher. Owls and hawks occasionally take an ermine, but one biologist has suggested that the black tail tip of this species limits raptor predation by focussing the predator's attention away from the weasel's vulnerable body. Humans trap thousands of ermine each season, although demand for their lush pelts has lessened in recent years. A parasitic nematode infects the nasal sinus of most *Mustela*, distorting and perforating the skull and possibly causing death. Most ermine live only 1–2 years, and 6–7 years may be the maximum.

Suggested References: King 1983 and 1990; Simms 1979b.

Long-tailed Weasel *Mustela frenata*

Measurements: Total length: 290–440 mm (11–17 in); tail length: 85–150 mm (3.3–5.9 in); hindfoot length: 30–50 mm (1.2–2.0 in); ear height: 15–23 mm (0.6–0.9 in); weight: 85–270 g (3.0–9.5 oz).

Description: The summer pelt is cinnamon brown above and yellowish white below. The base of the tail, outer legs, and usually the inner legs are also brown. In the northern Great Lakes basin, a long-tailed weasel takes on a white winter coat, but many individuals in the southern part of our area are brown year-round. The distal part of its tail is black at all times. This species is generally larger than the ermine, but a male ermine may be the same size as a female long-tailed weasel. However, the tail of a long-tailed weasel represents more than 44% of its head and body length, whereas the tail of an ermine is usually less than 40%.

Natural History: In the New World, this species has the most extensive distribution of any weasel and occurs from Canada to South America. In the Great Lakes drainage, a long-tailed weasel occupies forest-field edges, brushy fencelines, and wooded areas with shrubby cover. It often lives near water and can be found along marsh borders and in spruce or sphagnum bogs. The presence of humans, in moderation, does not bother

Long-tailed weasel. (Photo by Steven R. Sheffield.)

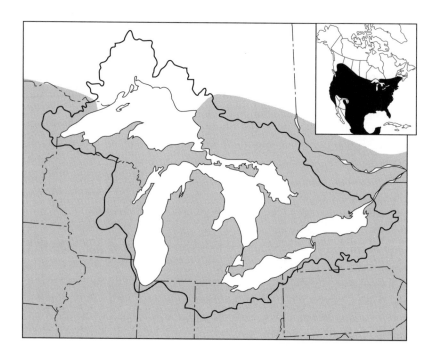

this carnivore; it lives about farm buildings and on the suburban fringe and, in general, seems more tolerant of disturbed areas than its cousin, the ermine. The long-tailed weasel is less abundant in the northern part of our area. Perhaps, as one biologist speculated, large males have difficulty running on snow and are just too big to follow prey through the subnivean environment.

As in the ermine, an individual long-tailed weasel patrols the boundary of its home territory and aggressively excludes others of the same sex. Although the larger home range of a male often overlaps that of one or more females, the sexes avoid each other outside the breeding season. This weasel is active throughout the year, mostly at night. It builds a grass- and fur-lined nest, generally in a mouse or mole burrow, but occasionally in a cavity under a building, log, or rock. The long-tailed weasel is not very tidy and often litters the nest and its surroundings with feces and food scraps.

It attacks prey with a lightning-fast rush, prevents escape by wrapping its slender body around the victim, and kills with a quick bite to the base of the skull. Cottontails, chipmunks, deer mice, jumping mice, meadow voles, and short-tailed shrews are typical prey. Birds make up 10% of the diet, and, to the dismay of many farmers, this weasel readily consumes chickens if natural prey are scarce. It fills out the menu with other vertebrates, insects, and berries. A long-tailed weasel consumes up to 70% of its body weight in food each day.

Mating takes place in midsummer, and birth occurs 8–10 months later, usually in March or April. An embryo spends most of this lengthy gestation in a quiescent state, and only during the last month does it attach to the uterine wall and actively develop. An average litter contains 4–8 pinkish young that are about the size of a small shrew; each weighs 3 g (0.1 oz) and has a total length of about 56 mm (2.2 in). At about three weeks of age, the canines and carnassial pair erupt, and a youngster begins eating meat supplied by its mother. The eyes open and weaning is completed during the fifth week of life. Shortly thereafter, it begins foraging with its mother and is capable of killing on its own by the eighth week. A young female mates when only 3–4 months old, but her male litter mates do not become sexually mature until the following summer.

Some long-tailed weasels survive at least three years in the wild, although little data exist on longevity. Mammalian predators include wild and domestic canids, bobcat, and lynx, and potential avian predators are large raptors, such as the great horned owl or rough-legged hawk. Many long-tailed weasels are infected with a nasal sinus nematode that may increase mortality. A few are killed by speeding automobiles, and trappers harvest many each year.

Suggested References: King 1990; Simms 1979a.

Least Weasel *Mustela nivalis*

Measurements: Total length: 170–210 mm (6.7–8.3 in); tail length: 22–44 mm (0.9–1.7 in); hindfoot length: 19–23 mm (0.7–0.9 in); ear height: 12–15 mm (0.5–0.6 in); weight: 30–60 g (1.1–2.1 oz).

Description: The least weasel is the smallest carnivore in the world, and its cylindrical body is only about 25 mm (1 in) in diameter. In the summer, the animal is a rich brown with whitish underparts. In response to the shorter days of autumn, a northern individual produces a completely white coat, but a least weasel in the southern Great Lakes basin retains its brown color throughout the year. The soles of the feet are furred year-round. The tail of a least weasel may have a few blackish hairs but never a distinct black tip, as in the ermine or long-tailed weasel. In addition, the tail is always less than 25% of head and body length in a least weasel but more than this in the other two species.

Natural History: The least weasel, like the ermine, ranges across northern Europe, Asia, and North America. It is a habitat generalist, occurring in fencerows, old fields, pastures, riparian edges, open woodlots, and, to a lesser extent, mature forests. A male and a female occupy overlapping home ranges, but each defends a central territory against others of the same sex. Home range size varies from 1 to 4 ha (2 to 10 acres) in females and 7 to 15 ha (17 to 37 acres) in males.

Small body size and an elongate shape promote rapid heat loss, and this energy must be replaced if the animal is to maintain a constant body temperature. Consequently, the least weasel is active both day and night in its search for food. During short inactive periods, it rests in a nest made from dried vegetation and insulated with a layer of fur or feathers; this soft lining is as much as 25 mm (1 in) thick in the winter. A favored nest site is the secluded burrow of a mouse or shrew that the weasel probably consumed as a recent meal.

Least weasel

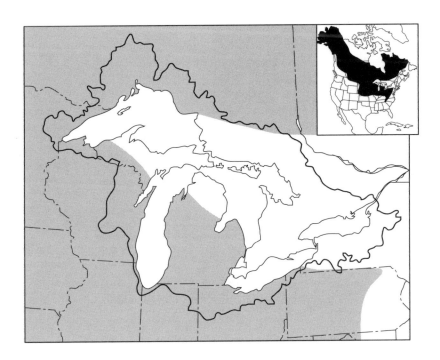

Mouse-sized mammals are the usual prey and include red-backed voles, southern bog lemmings, white-footed mice, masked shrews, and many others. Birds, amphibians, and insects contribute to the diet. Prey are encountered either above- or belowground, rushed, wrapped in a "weasel hug," and bitten repeatedly about the head and neck; a kill typically takes less than 60 seconds. This species quickly attacks moving prey but often ignores stationary animals, suggesting that it uses visual cues more than olfaction in detecting food. Compared to the smaller female, a male least weasel often consumes larger prey and forages more along the ground than in subterranean tunnels. Like other weasels, this species temporarily stores a surplus carcass in its nest tunnel or at a site closer to the kill.

Although the ecology of a least weasel is similar to that of the ermine and long-tailed weasel, its reproductive biology is very different. A least weasel is polyestrous, giving birth two or three times each year. Gestation is 35 days, and no delayed implantation occurs. Females typically give birth to five altricial young, each weighing only 1.5 g (0.05 oz), or less than a dime. Even though a young least weasel nibbles meat as early as 18 days of age, more than another week passes before the eyes finally open. By day 30, a youngster tags along on hunting trips, kills its first mice, and plays aggressively with siblings. It reaches adult size after 3 months of age and sexual maturity after 4–8 months.

Its tiny pelt has little market value, and the least weasel is not intensively sought by trappers. However, the small size of a least weasel makes it vulnerable to predation, and average life span is less than one year. Known mammalian predators include the long-tailed weasel, gray fox, red fox, and house cat. Least weasels are easily swallowed by snakes or carted away by hawks and owls.

Suggested Reference: Sheffield and King 1994.

Mink   *Mustela vison*

Measurements: Total length: 480–680 mm (19–27 in); tail length: 140–220 mm (5.5–8.7 in); hindfoot length: 50–75 mm (2.0–3.0 in); ear height: 15–25 mm (0.6–1.0 in); weight: 600–1,300 g (1.3–2.9 lb).

Mink

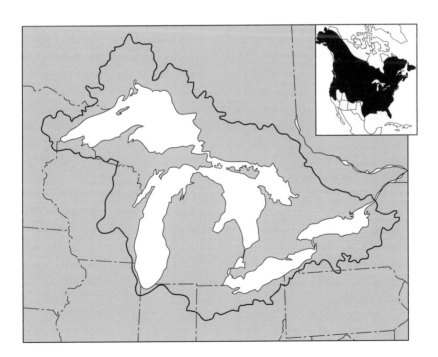

Description: The fur is soft, sleek, and thick. A mink generally appears deep brown throughout the year, except for occasional whitish blotches on the chin, throat, and/or belly. The toes are partly webbed and reflect this animal's semiaquatic life. Although a mink has the slender body typical of all *Mustela,* it is considerably larger than other members of this genus. It is similar in size and shape to a marten, but a mink has smaller ears, smaller feet, a less bushy tail, and a less prominent throat patch.

Natural History: The mink lives in most areas of Canada and the United States, except along the Arctic coast and in the arid Southwest. It occurs throughout the Great Lakes region. This common weasel frequents streams, ponds, and lakes with at least some brushy or rocky cover nearby.

When foraging, it generally stays within sight of open water, but it sometimes wanders up to 0.8 km (0.5 mi) onto dry land. Home range of a female is only 8–20 ha (19–50 acres), whereas that of a male is considerably larger. Like other weasels, each mink carefully marks the boundaries of its home using musky secretions from the enlarged anal glands. Although a mink vigorously defends its territory against others of the same sex, male and female home ranges often overlap.

A mink is active mostly at night, especially near dawn and dusk. It is a capable climber, travels quickly over land with a bounding gait, and is an excellent swimmer. While searching for food, this weasel cruises underwater for up to 30 m (100 ft) and dives to depths over 5 m (15 ft). It generally rests alone in an underground burrow that has an entrance close to the water's edge; this burrow is up to 3.5 m (12 ft) long, 30–90 cm (1–3 ft) belowground, and contains a nest chamber about 30 cm (1 ft) in diameter. A mink lines the nest cavity with dried grass and leaves, as well as with bits of fur plucked from its prey. This agile carnivore digs its own burrow or usurps the bank den of a recently deceased muskrat.

In the summer, the diet consists of crayfish supplemented mostly with frogs and small mammals, including shrews, rabbits, jumping mice, and muskrats. The mink also consistently eats fish, such as minnows and sunfish, as well as a variety of ducks and other fowl. In the winter, mammals, especially muskrats, are the primary food source. A hungry mink may attack these large rodents aboveground, in their den, or even underwater.

During late winter, a female mink enters heat and mates with one or more males. The gestation period varies from 40 to 75 days; it includes a developmental delay of 10–45 days, the length of which decreases as the mating season progresses. Females eventually give birth to a litter of 4–9 kits in April or May. Each six-gram (0.2-ounce) neonate appears wrinkled and pinkish but actually is covered by a coat of fine white hairs. After 5–6 weeks, the eyes finally open, and the kit begins supplementing its milk diet with meat provided by the mother. A juvenile makes its first kill between the 6th and 7th week, is self-sufficient by 8 weeks, and disperses from the natal den at 12 weeks. A female reaches physical maturity when 4 months old, mates at 10 months of age, and remains fertile for seven or more years. Males achieve physical and sexual maturity at 10 months.

Like the river otter, the mink accumulates mercury, polychlorinated biphenyls, and pesticide residues in its tissues, and these may affect its survival and reproduction. A few mink, especially dispersing youngsters, are killed by coyotes, red foxes, bobcats, and other large carnivores; some owls, such as the great horned owl, are also a threat. In reality, however, mink have few consistent enemies other than the fur trapper. Maximum longevity in captivity is 10 years.

Suggested Reference: Linscombe, Kinler, and Auerlich 1982.

American Badger *Taxidea taxus*

Measurements: Total length: 700–800 mm (28–31 in); tail length: 115–150 mm (4.5–5.9 in); hindfoot length: 95–140 mm (3.7–5.5 in); ear height: 45–60 mm (1.8–2.4 in); weight: 6–11 kg (13–24 lb).

Description: A badger resembles a very short-legged, medium-sized dog. The coarse dorsal fur is yellowish gray and very long, giving the animal a shaggy appearance. There is a white middorsal stripe extending from the nose onto the shoulders, and the whitish face is marred by a C-shaped black badge across each cheek. The claws on the forefeet are massive and much larger than those on the hindfeet.

Natural History: The badger dwells in open country and primarily resides in the western portion of the continent. It was not common or widespread in the Great Lakes basin before European settlement, but extensive forest clearing allowed this species to expand its range east-

American badger. (Photo by Bradley Fulk.)

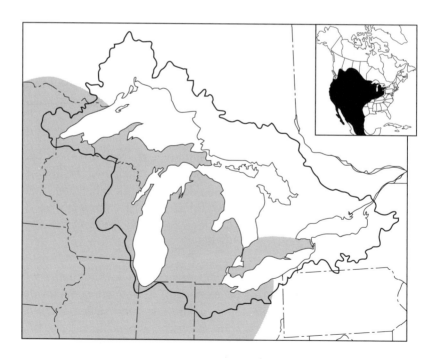

ward and northward into much of our area. It prefers grasslands, old fields, and pastures and rarely ventures into woody habitats.

As in other mustelids, there is little intrasexual tolerance, and males and females seemingly avoid each other for most of the year. Home range averages 240 ha (590 acres) for males and 160 ha (400 acres) for females; both sexes occupy smaller areas in winter compared to summer. A badger hunts mostly during darkness and spends the day in an underground den. Except for a nursing mother, this carnivore rarely occupies the same site on consecutive days during warm weather.

During the winter, a badger reduces aboveground activity by more than 90% and remains in its den for days at a time. The soil insulates the animal from external conditions so that, even if outside air temperature falls below -15°C (5°F), burrow temperature remains near 2°C (36°F). When at rest, a badger lowers its body temperature from a normal 38°C (100°F) to 29°C (84°F) for short intervals, averaging about 15 hours. Reduced activity, lessened exposure to harsh surface temperatures, and a small decrease in body temperature allow the badger to go for prolonged periods without eating. It survives the winter with occasional foraging and gradual depletion of body fat; fat represents more than 30% of a badger's body mass in November but less than 20% in March.

A badger's diet is heavily laden with fossorial rodents, such as the plains pocket gopher, thirteen-lined ground squirrel, and woodchuck. This lumbering carnivore does not win any speed or agility contests, but relies instead on its digging ability to obtain food. At the approach of a badger, a frightened prey retreats to the presumed safety of its burrow, but it is quickly dug out and consumed by the hungry carnivore. A badger also eats voles, mice, ground-nesting birds, bird eggs, grubs, caterpillars, and adult insects.

Mating occurs in August and September, but the blastocyst delays implantation until February. Ultimately, the female gives birth to 2–5 thinly furred youngsters in a grass-lined maternity den, sometime between late March and early May. Eyes open after 1 month, weaning occurs by 2–3 months, and juveniles disperse after 5–6 months. Although 30% of the females mate in their first summer, males do not breed until the following year.

A badger is fierce and powerful when cornered, and few, if any, animals are capable of successfully attacking it. Leading causes of death are automobiles, shooting, and fur trapping. Maximum longevity in confinement is 26 years. In the wild, the average life span is 4–5 years, and a few survive as long as 14 years.

Suggested References: Harlow 1981; Long 1973; Messick and Hornocker 1981.

Spotted Skunk *Spilogale putorius*

Measurements: Total length: 470–540 mm (18–21 in); tail length: 180–230 mm (7.1–9.1 in); hindfoot length: 45–50 mm (1.8–2.0 in); ear height: 25–30 mm (1.0–1.2 in); weight: 375–600 g (0.8–1.3 lb).

Description: Although black overall, this species has four or more white stripes along the back and sides that typically break into shorter bars or spots; no two spotted skunks have the same pattern. The head seems small compared to the body, and there is an inverted white triangle between the eyes. The bushy black tail is tipped with white. In comparison, a striped skunk is larger, has two stripes at most, no spots, and occasionally a narrow white line between the eyes.

Natural History: At one time, this western species ranged far into Wisconsin, to the edge of the Lake Michigan basin. However, it is now apparently extinct in that state and barely enters our region in northern Minnesota. It often lives in farmland, along brushy fencerows, streamside forests, and forest-field edges. The spotted skunk avoids the interior of dense forests as well as low-lying areas bordering ponds and marshes. In good habitat, density may be as high as 5 skunks/ha (2/acre).

During the breeding season, male home range is 500–1,000 ha (1,200–2,500 acres), but it is considerably less at other times of the year; a female restricts her activities to a smaller area than does a male. The den of a spotted skunk is a grass-lined cavity located in an abandoned gopher burrow, under a woodpile, beneath a farm building, or in a hollow log. Although basically terrestrial, the spotted skunk occasionally climbs trees to forage. It is active throughout the year, primarily at night. It is somewhat sociable in the winter, when up to eight skunks may occupy the same den.

In winter and spring, a spotted skunk preys heavily on small mammals, such as rabbits, voles, and mice. The summer and autumn diets, how-

Spotted skunk. (Photo by John and Gloria Tveten.)

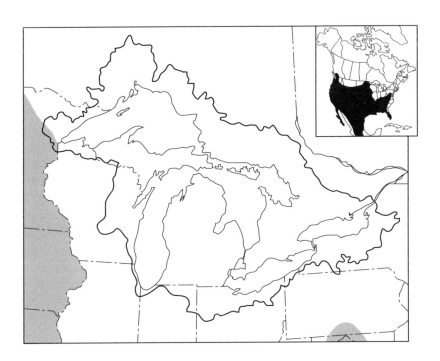

ever, contain more insect and plant material, especially fruits and grains. This mustelid often eats carrion, birds, and bird eggs, and occasionally attacks a snake or frog.

Reproductive biology apparently varies with geography. Western populations mate in autumn, exhibit delayed implantation, and have a gestation of 210–230 days. In contrast, spotted skunks in the Great Lakes drainage mate during April, experience no delay in development, and have a short gestation of 50–65 days. Litter size in our area varies from four to nine, and each newborn weighs about 10 g (0.3 oz). Young obtain the adult color pattern by day 21, and eyes open by day 32; weaning occurs after 54 days, and the pup approaches adult size after 100 days. When 9–10 months of age, a youngster mates for the first time.

This species possesses two large anal glands that secrete a milky-looking, foul-smelling musk. When mildly upset, a spotted skunk simply raises its white-tipped tail in warning. When sufficiently harassed, however, this mustelid indicates its displeasure by performing a handstand, while simultaneously straightening the tail, spreading its hindlegs, and arching its back until both head and anus face the enemy. Any intruder still foolish enough to approach within 2 m (6 ft) receives a well-aimed shower of fluid from the anal glands.

This spray is an effective deterrent against most mammalian carnivores, and they rarely prey on the spotted skunk. However, the sense of smell is weakly developed in birds, and large owls, such as the great horned owl, are a serious threat. Both striped and spotted skunks are prone to outbreaks of rabies, distemper, and other diseases. One spotted skunk lived for almost 10 years in captivity.

Suggested Reference: Howard and Marsh 1982.

Striped Skunk *Mephitis mephitis*

Measurements: Total length: 550–675 mm (22–27 in); tail length: 175–280 mm (6.9–11.0 in); hindfoot length: 60–80 mm (2.4–3.1 in); ear height: 25–35 mm (1.0–1.4 in); weight: 1.5–5.0 kg (3.3–11.0 lb).

Description: The striped skunk is mostly black but has a prominent white patch covering the top of the head and neck. Along the back and rump, this patch often splits into two broad stripes, although the back of some

Striped skunk

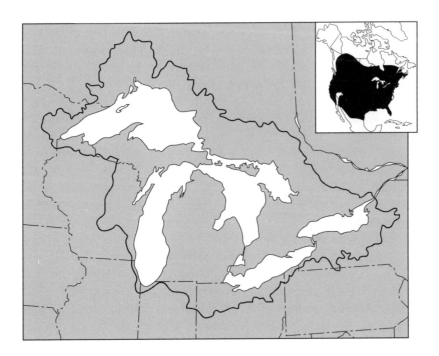

"striped" skunks may be totally black or completely white. This mammal has a small head, short legs, and a black, bushy tail that usually has at least a few white hairs at the end. Compared to the spotted skunk, the striped skunk is larger, has fewer stripes, and no spots.

Natural History: The striped skunk ranges from the Atlantic to the Pacific and from northern Mexico to Hudson Bay. This adaptable species is the common skunk of the Great Lakes region. Favored habitat is a mix of forests, fields, and wooded ravines, yet it also lives in areas of intense cultivation and suburban neighborhoods. It avoids dense stands of timber.

This docile mustelid generally dens in an abandoned woodchuck burrow located on a sunny hillside, but it sometimes rests under a building, brush pile, or jumble of rocks. Outside activity begins near dusk and continues until dawn. In this species, the home range often has an elongate shape and averages about 500 ha (1,200 acres) in males and 375 ha (900 acres) in females.

Winter is a difficult time for a striped skunk, and it uses a variety of strategies, similar to those of a badger, to survive the season. It remains in its den during cold snaps, and a dozen or more skunks may huddle together to conserve heat. While in the den, a skunk saves additional energy by lowering its body temperature, for short periods, from 38°C (100°F) to 32°C (90°F). Although it forages during mild weather, a striped skunk depends heavily on body fat to meet its energy demands, and weight loss over the winter may exceed 50%.

A striped skunk is an opportunistic feeder, and its diet changes with the season. During spring and summer, this carnivore is primarily insectivorous, consuming beetles, crickets, and grasshoppers. Later in the year, it adds corn and available fruits, such as mulberry, raspberry, black cherry, and grape. Other warm season foods include crayfish, worms, small mammals, birds, bird eggs, turtle eggs, and an occasional fish or frog. During the winter, it mostly feeds on small mammals, particularly the meadow vole.

Mating takes place in late March or April. Gestation varies from 59 to 77 days and includes a period of delayed implantation lasting up to 19 days. Ultimately, 4–6 sparsely furred young, weighing about 34 g (1.2 oz) each, are born in an underground nest. Even though eyes open after 3 weeks, youngsters do not venture far from the den until they are 6–7 weeks old. By this time, weaning has begun, and the curious kits line up behind their mother, like chicks behind a hen, and follow her on foraging expeditions. Juvenile males disperse in July or August, but females sometimes remain with the mother until spring. Both sexes mate just before their first birthday.

A striped skunk warns potential predators to stay back by performing a comical display. It raises its tail and arches the back, while repeatedly stomping its forelegs and shuffling backwards. If the foe is persistent, the skunk curls into a U-shape with both head and hindquarters facing the intruder. The skunk then sprays its oily musk in a wide arc, easily hitting targets up to 3 m (10 ft) away. The pungent fluid, which is a product of enlarged anal glands, often causes nausea and sometimes temporary blindness. A thorough dousing with tomato juice, followed by a conventional bath, helps eliminate the odor from a foolish pet or curious child.

So far, record longevity is 10 years in captivity and at least 4 years in the wild. Although mammalian carnivores generally avoid skunks, large birds of prey, such as the great horned owl and golden eagle, are seemingly not affected by the defensive musk. Humans trap a small number of striped skunks for their fur each year and crush many more with their automobiles, especially during the late-winter mating season. Many striped and spotted skunks are parasitized by a frontal sinus nematode that erodes, perforates, and warps the frontal bone of the skull; this deformity may increase pressure on the brain and ultimately increase mortality.

Suggested References: Maldonado and Kirkland 1986; Wade-Smith and Verts 1982.

Northern River Otter   *Lutra canadensis*

Measurements: Total length: 900–1,300 mm (35–51 in); tail length: 320–500 mm (13–20 in); hindfoot length: 110–135 mm (4.3–5.3 in); ear height: 10–25 mm (0.4–1.0 in); weight: 5–14 kg (11–31 lb).

Description: The sleek, waterproof fur of a river otter is a rich brown along the back, grayish brown below, and has a touch of silver about the throat and muzzle. Ears are inconspicuous, but the vibrissae (whiskers) are very long. The thick, muscular tail tapers to a point and represents at least one-third of the total length. As an adaptation to its semiaquatic life,

Northern river otter. (Photo by the author.)

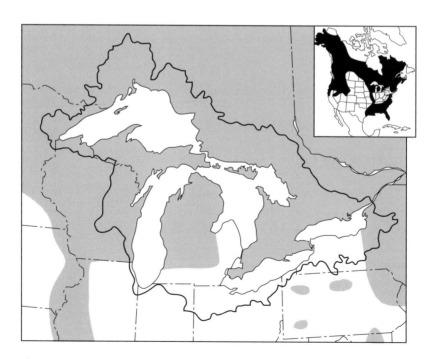

a river otter has five fully webbed toes on each foot. Although this species has the slender body and short legs typical of its family, a river otter is larger than any other mustelid currently living in the Great Lakes basin except the badger.

Natural History: The original distribution of the river otter included most of Canada and the United States, although overtrapping, urbanization, and pollution had greatly reduced its range by the early twentieth century. Fortunately, the otter is making a steady comeback, slowly repopulating some areas on its own and being reintroduced into others by conservation agencies. Above all else, this graceful mammal requires clean, moderately deep water in the form of streams, ponds, or lakes. A river otter has a penchant for waterways with steep banks, but the surrounding terrestrial vegetation does not greatly influence otter abundance.

Home is usually a subterranean nest chamber connected by an underwater entrance to an adjacent stream or pond. A river otter rarely is the original excavator; instead, it remodels a bank den previously dug by a muskrat or beaver. Other potential resting places are among tree roots, in a hollow log, or inside a beaver lodge.

An otter forages primarily at night, throughout the year, on land or in water. It swims at speeds up to 11 km/hour (7 mi/hour), faster than many fish that it preys on, and dives to depths approaching 18 m (60 ft). On land, this playful creature runs with an unspectacular loping gait. However, if snow or ice is available, the otter takes about three strides, jumps head-first, and begins a wobbly slide on its chest and belly; such slides often cover 3.0 m (10 ft) on snow and up to 7.5 m (25 ft) on ice. A river otter also "toboggans" on muddy banks and over grassy vegetation.

A river otter is a fish specialist, concentrating on minnows, carp, suckers, darters, and sticklebacks. It occasionally takes faster-moving prey, such as trout, perch, and sunfish, but the otter is not a serious threat to gamefish. Prey are captured with a quick lunge from ambush or, more rarely, after a prolonged chase. Crayfish, clams, amphibians, and small mammals add variety to the menu.

Intrauterine life varies from 270 to 380 days and includes a 9–11 month period of suspended development. Most births occur during late March or April, and the new mother quickly enters heat, copulates, and begins another prolonged gestation. Litters usually contain two or three young. Otter kits are about 28 cm (11 in) long and weigh 130 g (4.6 oz) at birth; they are fully furred but toothless and blind. Eyes open after 5 weeks. Kits begin taking solid food at 9–10 weeks, and they are totally weaned 2 weeks later. The mother supervises the first family swim at around 8 weeks of age and later brings live prey to the youngsters so that they can practice their hunting skills. Although some families disband in autumn, others remain together until spring. Females successfully mate near their second birthday, but males generally wait until they are five years old.

The major threats to this species are unregulated fur trapping and habitat degradation. A river otter is the top link in an aquatic food chain, and it readily accumulates dangerous levels of mercury and organo-chlorine compounds. Maximum longevity is 13 years in the wild and 20 years in captivity.

Suggested References: Beckel 1991; Melquist and Hornocker 1983; Ropek and Neely 1993.

Cats
Family Felidae

Many of the most feared predators in the world, both present and past, belong to this highly specialized family. Extinct saber-toothed cats that sport canines as long as the entire skull belong to the Felidae, as do modern leopards, lions, and tigers. Felids first appear in the fossil record during the early Eocene epoch, over 50 million years ago. Today, there are 36 different species arranged into 3–15 genera, depending on which authority one consults. This family occurs naturally in most parts of the world except Antarctica, Australia, and a number of large oceanic islands, such as New Zealand and New Guinea. Although two species (lynx and bobcat) still live in the Great Lakes region, a third (mountain lion) disappeared by early in the twentieth century.

Wild felids range in size from the African black-footed cat at 1.5 kg (3.3 lb) to the magnificent tiger that weighs as much as 380 kg (840 lb) and stretches 3.5 m (11.5 ft) from nose to tip of tail. Despite the size variation, all cats have the same basic body plan and general habits. Cats, for example, have four toes on the hindfeet and five on the front, although the pollex (thumb) is small and generally does not contact the substrate. All members of the family are good runners with a digitigrade stance. During most activities, the last segment of each toe and its attached claw are rotated up and back, or retracted, thus preventing the claw from becoming dulled by repeated contact with the ground. During prey capture or tree climbing, however, the sharp claws are rotated outward. In addition, a cat has a long, mobile tongue, covered with stiff spines that are used to remove even small bits of flesh from a bony carcass. Finally, their excellent vision, hearing, and olfaction make these carnivores superb hunters.

The skull also reflects a predatory way of life (fig. 19). The long, curved canines are used to hold and puncture prey, whereas the well-developed carnassial pair slices the victim into pieces. The rostrum is short and provides for a strong bite. A curved profile to the skull, 28 or 30 teeth, and small last upper molar are typical of felids.

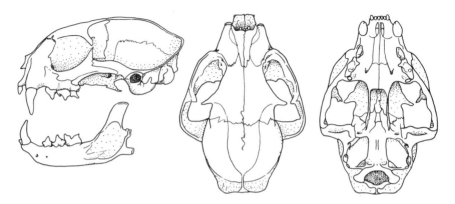

Fig. 19. Skull of a bobcat, family Felidae

Canadian Lynx   *Lynx canadensis*

Measurements: Total length: 825–1,050 mm (32–41 in); tail length: 75–125 mm (3.0–4.9 in); hindfoot length: 205–250 mm (8.1–9.8 in); ear height: 70–80 mm (2.8–3.1 in); weight: 5–17 kg (11–37 lb).

Description: The lynx is a beautiful, short-tailed cat with a black-striped ruff running down the cheeks. Its long, thick, dorsal fur is usually yellowish brown, and it sometimes has a gray frosted appearance; spotting is variable and generally more evident on the sides and legs. The ear is capped by a black tuft of hairs that is as long as 50% of the pinna height. The feet are very broad, and the tail tip is black all around. The lynx is similar in size and appearance to a bobcat; however, a bobcat has smaller feet and shorter ear tufts, and its tail tip has black markings only on the dorsal surface.

Natural History: This species occupies the northern part of North America, and some mammalogists consider it to be conspecific with the Eurasian cat *Lynx lynx*. The lynx occupies mature coniferous forests that are strewn with thick litter, ferns, and rotting logs. This cat is at home in cedar swamps, as well as upland woods of hemlock and fir. It avoids disturbed areas. At one time, it lived throughout the Great Lakes basin, but human activities forced the lynx into a slow northward retreat, into uncut boreal timber. Although a few cats occasionally are sighted in Michigan and

Canadian lynx. (Photo by Lawrence L. Master.)

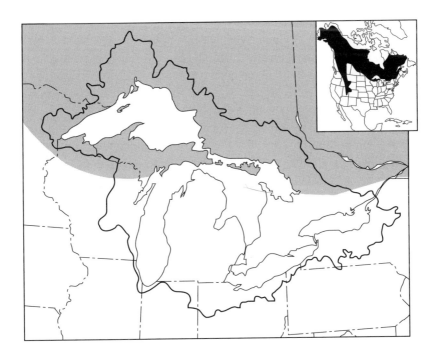

Wisconsin, their numbers are quite low, and there is little evidence of breeding activity.

A lynx is a solitary hunter, prowling about a spacious territory of 10–50 km² (4–20 mi²). The resident cat scent-marks its boundaries with urine and feces, and, unlike the bobcat, individuals of both sexes avoid trespassing. Although a lynx often lives its entire life within a single well-defined territory, food shortages sometimes induce migratory movements up to 400 km (250 mi). A lynx forages mostly at night and rests during the day under a rocky overhang or downed tree. This animal climbs well, and its broad feet allow it to move easily over crusted snow.

A lynx stalks or ambushes prey and seldom chases its dinner for more than 50 m (165 ft). Snowshoe hares are preferred fare, forming 65–85% of the diet. The lynx is so dependent on hares that the cat's population fluctuates in synchrony with that of the hares and exhibits the same 10-year cycle of abundance. Other mammals commonly eaten by this cat include red squirrels, meadow voles, and muskrats. Although the lynx preys on white-tailed deer fawns, it rarely attacks an adult. Carrion, ducks, and upland birds, such as ruffed grouse, complete the diet.

Mating activity peaks in March, and a litter of 2–5 young is born 63–70 days later. The toothless newborn are well furred, weigh about 200 g (7 oz) each, and have a total length of 170 mm (7 in). Eyes open after 14 days. The kittens take their first tentative steps by day 26 and begin playful stalking by day 40. Youngsters consume prey killed by their mother starting at 1 month of age but continue to suckle for another 3–5 months. A few females breed when 10 months old; however, most female and all male lynx wait until at least their second year.

Although most perish before their fourth or fifth birthday in nature, a few live as long as 15 years. Humans are a continual threat through hunting and trapping, but most mortality, especially among young lynx, is related to food shortages. Lynx have survived for 24 years in a zoo.

Suggested Reference: Tumlinson 1987.

Bobcat   *Lynx rufus*

Measurements: Total length: 750–1,100 mm (30–43 in); tail length: 130–180 mm (5.1–7.1 in); hindfoot length: 160–195 mm (6.3–7.7 in); ear height: 60–85 mm (2.4–3.3 in); weight: 5–16 kg (11–35 lb).

Bobcat

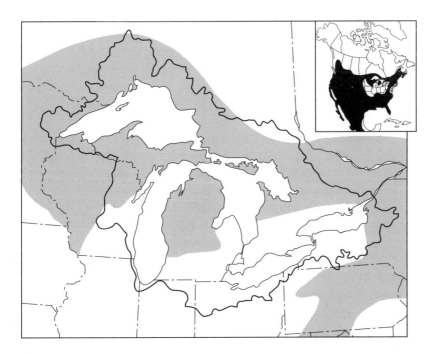

Description: Even though the bobcat is the smallest of our wild cats, it is 2–3 times larger than a typical house cat. Overall, it is grayish to reddish brown along the back and white underneath; irregular spots or blotches occur on both dorsal and ventral surfaces. It has a prominent ruff running along the cheeks and giving the appearance of a furry jowl. The short tail has several dark bars near the tip that do not continue onto the underside of the tail, unlike the single circular band found in the lynx. Ear tufts are present but much less pronounced than those of a lynx.

Natural History: The bobcat lives along the southern fringe of Canada, throughout most of the United States, and far into Mexico. Although this cat once roamed the entire Great Lakes basin, it is now generally absent in southern areas dominated by cities and endless farmland. Nevertheless, individuals occasionally appear far outside of the range indicated; a male bobcat, for example, was trapped in extreme northeastern Indiana in 1993. Preferred habitat is large tracts of hardwood forest, but it also lives in coniferous and mixed deciduous/coniferous woods. It readily occupies wooded swamps and often stays close to riparian forests in areas thickly populated with humans.

A bobcat is a loner; it marks its territory with urine and feces and excludes others of the same sex. Scratching posts—trees scarred by this cat while sharpening its claws—also mark the animal's home. Depending on prey availability, home range varies from 10 to 100 km^2 (5 to 40 mi^2), and an individual covers as much as 12 km (7 mi) in its nightly search for food. Except for a nursing mother, this cat changes resting spots almost daily, retreating under a rocky ledge, beneath tangled brush, or into a tree cavity.

The spotted coat provides this stealthy hunter with excellent camouflage. It often stalks its prey, breaks from cover with a quick rush, pounces, and makes the kill. Alternatively, it lies next to a game trail or sits motionlessly in a tree and ambushes any animal that wanders by. Eastern cottontails and snowshoe hares are common prey, but a bobcat eats most small mammals that are present in our northern forests; shrews, voles, mice, jumping mice, bog lemmings, squirrels, opossums, and even porcupines are reported foods. This cat readily feasts on deer carrion, and it sometimes preys on sick or young deer as well. An occasional frog, snake, or bird adds variety to a diet dominated by mammals.

A typical litter contains 1–4 kittens that are born in May after a gestation of about 60 days. A bobcat begins life with a full set of claws, closed eyes, and a woolly coat of spotted fur; it weighs 325 g (11 oz), the size of a small gray squirrel, and is 250 mm (10 in) long. Immediately after birth, a kitten nurses while the mother licks it repeatedly, cleaning and drying the fur. She eventually eats the placenta and stays with the newborn for two days before leaving to hunt.

Growth is not very fast, and well-fed youngsters gain only about 10 g/day (0.4 oz/day). Eyes open by day 10. After one month, a kitten begins nibbling meat brought by the mother, but weaning is a slow process that is not complete for another month. The mother supervises the kittens until midwinter, when she drives them off and prepares to mate again. Most bobcats breed at the end of their second year. If an adult loses her litter early in the season, she comes into heat again and gives birth as late as September.

Although kittens occasionally are killed by foxes, coyotes, or large owls, an adult bobcat is safe from predation. Nevertheless, there still is a market for the bobcat's spotted pelt, and a few thousand of these cats are hunted or trapped each year in the Great Lakes region. Wild bobcats live up to 12 years, and those in captivity, up to 32 years.

Suggested Reference: McCord and Cardoza 1982.

Even-toed Ungulates
Order Artiodactyla

This order includes familiar domesticated animals, such as goats and cattle, and a multitude of wild mammals, including peccaries, tapirs, bison, and deer. The smallest artiodactyl is the Asiatic mouse deer that weighs as little as 1 kg (2.2 lb) and stands only 250 mm (10 in) at the shoulder. In contrast, the African giraffe is the tallest of all mammals, with an average shoulder height of 3 m (10 ft) and total height of 5 m (17 ft). The ponderous hippopotamus weighs up to 4,500 kg (10,000 lb) and is one of the heaviest of land mammals; just one if its canine teeth weighs more than an entire mouse deer!

An ungulate is a mammal that has hooves instead of nails or claws. The artiodactyls are even-toed because they have either two or four functional toes on each foot; this is in contrast to the order Perissodactyla, or odd-toed ungulates, that generally have one (e.g., horse) or three (e.g., most rhinoceroses) useful digits. Almost all ungulates have an unguligrade stance, that is, the only part of the limb to contact the ground is the distal tip of the toe and its surrounding hoof. This is an evolutionary adaptation that effectively lengthens the leg, allowing the animal to take longer strides and, ultimately, run faster than more flat-footed mammals.

The two largest families in the Artiodactyla are well known for their head ornaments. True horns are found in the cow family (Bovidae), whereas antlers characterize members of the deer family (Cervidae). Bovid horns are living, bony extensions of the skull surrounded by a nonliving keratinized sheath; they are never branched and may be found in both males and females, depending on species. Antlers are also bony outgrowths of the skull, but they lack the surrounding sheath, often are branched, and, in all but one species, are restricted to males. Horns are permanent structures, whereas antlers are shed annually.

Most artiodactyls consume low-quality foods, such as grasses, forbs, or twigs. To digest such a coarse diet, these mammals generally have a three- or four-chambered stomach, including a rumen that houses a teeming colony of microorganisms. The animal briefly chews its food and then passes it to the rumen where microbes begin breaking down cellulose, the main constituent of plant cell walls. After a time, the partly digested food (cud) is regurgitated, chewed again, and reswallowed. This process slowly, but efficiently, breaks down fibrous plant material by alternating the grinding action of teeth and chemical action within the rumen.

The natural distribution of artiodactyls is virtually worldwide except isolated oceanic islands, Antarctica, and marsupial-dominated Australia. They dwell in virtually all habitats from sea level to high mountains, from the tropics to the Arctic. Currently, taxonomists recognize 9 living families with 81 genera and over 200 species. Only one family, the Cervidae, is present in the Great Lakes region, although the Bovidae, represented by the American bison, lived here until the late eighteenth century.

Deer
Family Cervidae

This family is native to North and South America, Asia, Europe, and northern Africa, and it has been introduced into many areas, including Australia, New Zealand, New Guinea, and Cuba. Cervids occupy a variety of habitats, ranging from dense forests to open prairies, tropical jungles to Arctic tundras, and arid savannas to temperate swamps. Humans hunt many deer species for meat and fur and have domesticated one species, the reindeer (caribou).

The largest member of the family is the North American moose, which weighs up to 800 kg (1,760 lb), whereas the smallest of the true deer is the South American pudu at only 6 kg (13 lb). Males are generally larger than females. Like other artiodactyls, cervids have long legs and walk on their toe tips. They are excellent runners and jumpers, and some are capable of reaching speeds of 70 km/hour (42 mi/hour) and hurdling obstacles that are 3 m (10 ft) high.

The skull and teeth differ from those of the other mammals considered so far. Upper incisors are never present, upper canines are absent in most, and lower canines are incisorlike (fig. 20). When feeding, a cervid holds a plant between its lower incisors and upper jaw and tears off the desired piece with a slight head jerk—a process analogous to ripping cellophane tape from a dispenser. The high-crowned cheek teeth have crescent-shaped ridges for grinding fresh food as well as regurgitated cud. The skull is fenestrated anterior to the eye orbit, and the premaxillary bones project much farther forward than the short nasal bones.

In addition, antlers often are present. These structures are bony outgrowths of the frontal bone that function in defense and sexual advertisement. They are replaced annually. In white-tailed deer, for example, antler growth occurs between May and September, and shedding takes place during midwinter. Throughout spring and summer, a developing antler is nourished by "velvet," a living tissue containing nerves and blood vessels and coated with fine hairs. When growth is complete, the velvet dries, and the buck removes it by rubbing against tree branches. The growth process is under hormonal control and regulated by changing photoperiod. Growth rate, as well as the ultimate size and complexity of antlers, is a function of maturity and diet, and deer that are under nutritional stress often produce a stunted rack. Only male cervids grow antlers, except caribou, in which both sexes may possess them.

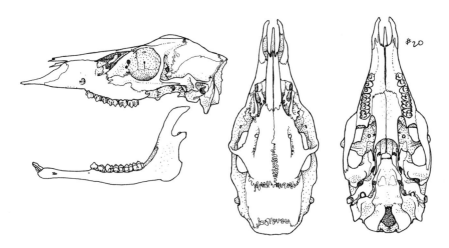

Fig. 20. Skull of a white-tailed deer, family Cervidae

Fossils from Asia indicate that cervids existed at least 35 million years ago, during the early Oligocene epoch. Today, there are 16 genera and 43 living species of deer worldwide. In the Great Lakes region, there are four species, each representing a different genus. Two of these, white-tailed deer and moose, are widespread, whereas caribou live only in the extreme north. Elk disappeared from our area before 1900, but humans have successfully reintroduced the species to isolated areas.

Wapiti or Elk   *Cervus elaphus*

Measurements: Total length: 2,100–2,700 mm (6.5–9.0 ft); tail length: 120–160 mm (4.7–6.3 in); hindfoot length: 600–655 mm (24–26 in); ear height: 140–200 mm (5.5–7.9 in); weight: 240–450 kg (530–1,000 lb).

Description: Elk are grayish brown on the back and sides, and darker on the head, neck, legs, and belly. There is a distinct buff-colored rump patch surrounding a short white tail and a dark shaggy mane hanging from the neck and chest. An antler consists of a single cylindrical beam that curves up and back over the shoulders and bears several tines; the beam of mature individuals is 1.1–1.6 m (3.5–5.0 ft) in length, as is the spread. Both male and female elk have upper canines.

Elk. (Photo from the Roger W. Barbour Collection, Morehead State University.)

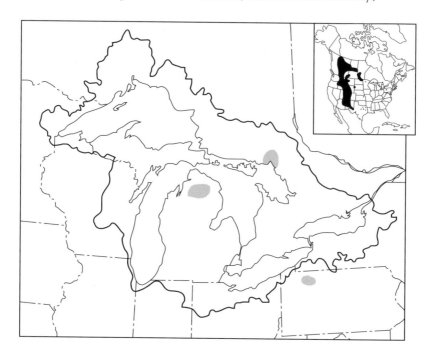

Natural History: Elk live in both the Old World and New World. This magnificent animal once roamed over most of the United States and Canada, but incessant hunting and habitat destruction severely reduced its range. It disappeared from Ontario as early as 1825 and from Wisconsin, Pennsylvania, and Michigan by 1880. Beginning early in the twentieth century, wild stock from western North America was released at various sites in the East. Today, the only substantial populations actually within the Great Lakes basin are in northern Lower Michigan and in the Burwash–French River area of Ontario. This species generally avoids dense unbroken forests, preferring open meadows or woodlands interspersed with grassy clearings.

Elk are gregarious for most of the year, but the type of social unit varies with the season. Cows and their offspring spend much of the summer in bands of 5–30 individuals, while males are either solitary or form small bachelor groups. During September, fierce fighting occurs among bulls, and eventually the largest and strongest males obtain a harem of 20 or fewer cows. A bull tightly controls the movements of harem members and services each female as she comes into heat. After the mating season, harems break up, and many elk congregate in larger winter herds; some adult males join these herds, but other bulls return to a solitary existence.

An elk feeds mainly at night and near dawn and dusk; during midday, it beds down and chews its cud. In spring and summer, it grazes on tender young grasses and various forbs, such as dandelion, aster, hawkweed, violet, and clover. Mushrooms are relished. With the coming of autumn, an elk becomes more of a browser, taking twigs and bark from white cedar, wintergreen, hemlock, sumac, maple, and basswood. An elk, however, prefers to graze rather than browse, and even during midwinter, it seeks out windswept patches or paws through crusted snow to reach the grass below.

In September, antler growth is complete. A bull then attracts females with a piercing bugle and vigorously defends his harem with threats, body shoves, and antler stabs. Mating takes place at this time and birth occurs in May or June, after a gestation of about 250 days. The single calf weighs 16 kg (35 lb), has a spotted coat, and follows its mother within just a few days. Most young are weaned by September but stay with the cow throughout the winter. Although some females mate when 16 months old, most wait until the following year. Males have no hope of defending a harem until they reach adult size, after their fourth or fifth birthday.

264 / Mammals of the Great Lakes Region

Maximum longevity in a zoo exceeds 20 years, but most individuals in the wild live less than 10 years. Males experience higher mortality than females because of injuries received during the rutting season and extensive harvesting by hunters. Fawns and old adults occasionally fall prey to the gray wolf, coyote, lynx, bobcat, and black bear. In addition, the meningeal worm, a parasitic nematode that infects tissues surrounding the brain, causes some mortality. Poaching may be the most serious threat to the small elk herds of the Great Lakes region.

Suggested References: Beyer 1987; Thomas and Toweill 1982; Ranta, Merriam, and Wegner 1982.

White-tailed Deer   *Odocoileus virginianus*

Measurements: Total length: 1,550–2,100 mm (5–7 ft); tail length: 250–350 mm (10–14 in); hindfoot length: 475–525 mm (19–21 in); ear height: 140–225 mm (5.5–8.9 in); weight: 45–140 kg (100–300 lb).

Description: This small cervid has a reddish brown coat in summer and appears more grayish in winter. The belly and underside of the tail are snow white. There is a white throat patch, white eye ring, and narrow white band behind the black nostrils. The major beam of the antler bends forward.

Natural History: The white-tailed deer is a wide-ranging mammal that occurs from South America to central Canada. In our area, this species prefers open forest environments interspersed with meadows, woodland clearings, or farmland. Before European settlement, the northern Great Lakes region was covered with climax coniferous forest and provided very poor habitat for this mammal, but extensive logging in the late nineteenth century opened the forests and resulted in a huge increase in the deer population. At the same time that the northern herd was expanding, southern populations were decimated through overhunting. However, government control of the hunt and abundant food from productive farms have allowed these southern deer to recover to the point that some consider them pests.

The white-tailed deer is a little less social than the elk. An adult female lives much of the year in an extended family group with her newest fawns

White-tailed deer

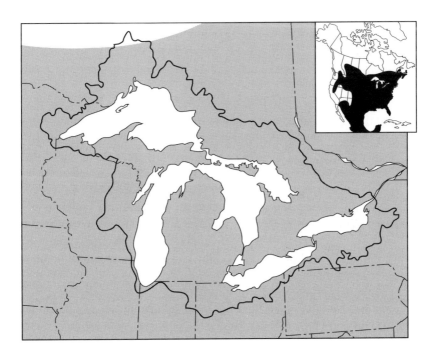

and her female yearling offspring. A male, in contrast, is solitary or joins a small bachelor group.

In northern areas with heavy snow, deer spend the winter at traditional locations in groups numbering up to 50 animals. Such "yards" are often in low-lying areas with dense cover. Here, the deer take a sit-and-wait attitude, reducing activity levels and energy demands, but slowly depleting the local food supply, as well as their small store of body fat. If the winter is short and mild, most survive; if it is long and cold, many starve.

This deer forages mostly in the dim light of sunrise and sunset but is active at any time of day or night. It feeds while slowly walking, never staying in one place for long. In the winter, a white-tail browses on buds and twigs nipped from maple, dogwood, aspen, willow, and sumac; in the north, a deer depends more on evergreens, particularly white cedar. In the summer, it munches on grasses, herbs, and leaves. The white-tail relishes acorns and mushrooms and competes with man for apples, corn, celery, and soybeans.

The rutting season lasts for two months in autumn and peaks in November. During this time, a male wanders the countryside, searching for mates and threatening any other male that stands in his way. Occasionally, two bucks battle each other with antlers and hooves until one withdraws, leaving the other free to associate with any available doe. The victor may then follow a prospective mate for days, waiting for her to come into heat, yet he quickly leaves after copulation in search of another. A doe that is not successfully fertilized has a second period of heat 28 days later.

In late May or June, fawns weighing 2–3 kg (4–7 lb) are born, about 200 days after conception. Mature does usually produce twins, but yearling females have a single offspring. Although a young white-tail stands shortly after birth, it does not follow the mother as she forages; instead, the doe caches the white-spotted fawn in a secluded glade and returns periodically to nurse. After a few weeks, the youngster begins nibbling grass and follows the parent wherever she goes. Weaning is a slow process and is not completed until the fawn is four months old and has shed its spotted coat. A few individuals become sexually mature during their first winter, but a male has little chance of mating until he becomes older and stronger.

Although predation of healthy adults is rare, the coyote, gray wolf, lynx, bobcat, and black bear occasionally take fawns and infirm adults. Feral dogs are probably the biggest predation threat in much of the Great

Lakes basin. Hunters, of course, bag many of these animals for table meat, and thousands die after colliding with automobiles; deer-auto collisions are most common in November when the rut peaks and deer lose much of their normal cautiousness. White-tailed deer live up to 20 years in captivity and 15 years in nature; wild individuals seldom survive more than 3–6 years.

Suggested Reference: Smith 1991.

Moose *Alces alces*

Measurements: Total length: 2,000–2,800 mm (6.5–9.0 ft); tail length: 80–120 mm (3.1–4.7 in); hindfoot length: 730–835 mm (29–33 in); ear height: 255–265 mm (10 in); weight: 330–500 kg (725–1,100 lb), occasionally larger.

Description: This horse-sized animal is the largest of all cervids. It is grayish or reddish brown or occasionally black. The muzzle is broad with an overhanging snout, and the shoulders are distinctly humped. A flap of skin, the "bell," dangles from the neck. A large head, small rump, and long, spindly legs contribute to an overall awkward appearance. The main beam of the massive antler extends laterally, flattens out, and curves backward.

Natural History: The moose is Holarctic in distribution, ranging across northern Europe, Asia, and North America. It is a resident of boreal forests, particularly early successional stages dominated by shrubby growth and immature trees. Especially in summer, it frequents such moist habitats as cedar swamps, marshes, and alder-willow thickets bordering waterways. Home range is 2–20 km^2 (1–8 mi^2) and is up to five times larger in summer compared to winter. Typical density is one moose every 2 km^2 (5/mi^2).

A moose spends most of the day bedded down, chewing its cud, and forages primarily at dawn and dusk. Although moose sometime congregate at favored feeding sites, individuals ignore each other outside the breeding season. Despite its ungainly appearance, a moose is an excellent swimmer and, when motivated, can ramble along on land at speeds approaching 55 km/hour (35 mi/hour). This cervid has keen hearing and an acute sense of smell but poor vision.

Moose. (Photo by John and Gloria Tveten.)

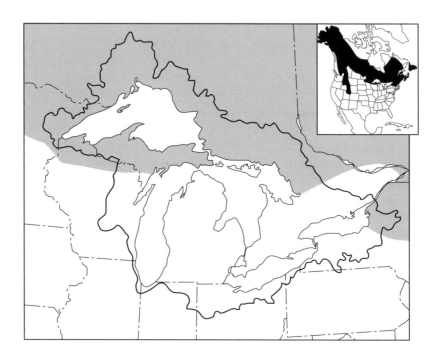

Unlike white-tailed deer and elk, a moose rarely grazes on grass. In late spring, it feeds on tender young leaves ripped from trees, such as aspen, maple, and birch, as well as shrubby willows, alders, and dogwoods. By summer, it shifts its attention to lakes and rivers, where it spends hours uprooting plant after plant. It is fond of water lilies but also eats rushes, arrowheads, horsetails, and aquatic sedges. A moose may wade far from shore and even temporarily submerge itself during its quest for the roots and basal stems of some plants. The winter diet is mostly twigs and buds. A hungry moose requires about 20 kg (44 lb) of forage each day.

The rut begins in September and continues into October. At this time, each male is extremely restless and pugnacious, constantly searching for females and challenging other bulls. A male moose may associate with a small group of females at this time, but he does not jealously control the movements of the cows; they are free to come and go. Many bulls simply stay with a single female for a few days until she permits copulation, after which he begins a feverish search for a new mate.

Gestation lasts 240 days, and one or two young are born in late May or early June. At birth, a moose calf is cloaked in reddish brown fur, without spots, and weighs 11–16 kg (25–35 lb). A newborn sees and hears quite well but has trouble standing for any length of time; however, it is steady on its feet after 5 days and eagerly follows the cow after 15–20 days. The calf continues to suckle for five months and remains with its mother until she is ready to give birth again the following spring. Most females do not mate until they are in their third year, and many continue to do so until 18 years old. The majority of males do not breed until their sixth year.

In white-tailed deer, the meningeal worm is a common parasite that lodges in the membranes covering the brain (the meninges) but does little apparent harm to the deer. During the past two centuries, logging operations opened the boreal forests and allowed white-tailed deer to spread northward into the range of the moose, bringing this parasitic nematode along with it. Unfortunately, the worm may cause paralysis and eventual death in moose, and some wildlife biologists believe that the parasite now is a major mortality factor for moose in the Great Lakes basin. In terms of predation, only the gray wolf has a chance at bringing down a mature individual, but wild cats and black bears frequently attack calves.

Maximum longevity is 27 years, which is similar to some large carnivores as well as our tiny bats.

Suggested References: Boutin 1992; Franzmann 1981; Nudds 1990.

Caribou or Reindeer   *Rangifer tarandus*

Measurements: Total length: 1,650–2,300 mm (5.5–7.5 ft); tail length: 100–150 mm (3.9–5.9 in); hindfoot length: 500–660 mm (20–26 in); ear height: 110–150 mm (4.3–5.9 in); weight: 90–180 kg (200–400 lb), occasionally larger.

Description: Caribou are various shades of brown with a darker face and chest. The neck and mane are yellowish white, whereas the belly and rump are pure white. The brown leg terminates in an extremely broad foot with two large dewclaws, and there is a touch of white just above the hooves. Unlike other cervids, both male and female caribou sport antlers, although the frequency of antlered females differs between populations and over time. The main beam curves back, up, and forward, and there often is a prominent, flaring, brow tine that projects down and over the forehead. The right and left antlers are rarely symmetrical in caribou.

Natural History: The caribou ranges from the upper Great Lakes to far north of the Arctic Circle; it also occurs in Europe and Asia, where it is better known as the reindeer. Although many people associate this mammal with long treks across barren tundra, caribou in the Great Lakes region are forest-dwellers. During postglacial times, this species occurred as far south as Detroit and was present in Wisconsin and Michigan as late as 1910 and Minnesota until the 1930s. Today a few still live in dense coniferous woods along the Canadian shore of Lake Superior and occasionally migrate into extreme northern Minnesota during harsh winters.

Like other cervids, it feeds primarily near sunrise and sunset and spends the remaining hours chewing its cud. A caribou walks continually while feeding and, when frightened, can reach speeds of 65 km/hour (40 mi/hour); it is a strong swimmer and often fords large rivers during its seasonal wanderings. It depends heavily on a keen sense of smell to detect predators and locate food under snow, but its hearing and vision

Caribou

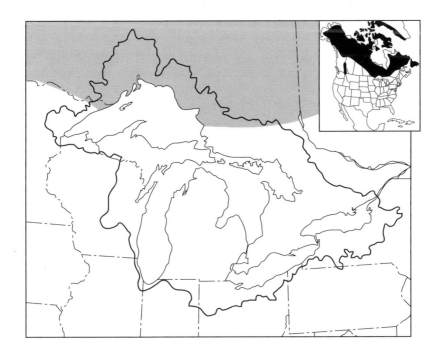

are average, at best. Woodland caribou are less migratory and less social than tundra caribou; the forest form gathers in small groups of less than 10 animals and has a home range of 50–200 km^2 (20–75 mi^2).

A caribou's diet varies with the season. During spring and summer, it nibbles mushrooms, grasses, herbs, and fruits, and browses on leaves from birch and willow. When snow covers the ground, a caribou concentrates on twigs and winter buds of evergreen trees and shrubs. A caribou also makes use of an unusual food source—lichens; it consumes lichens year-round and eats those found on the ground as well as those clinging to the trunks and branches of conifer trees.

During October and November, a bull guards a small harem and inseminates each cow as she comes into heat. The single calf is dropped 228 days later, in late May or early June. The mother removes the afterbirth, thoroughly licks her newborn, and, for the first few hours, keeps it warm by lying next to it. Neonates are clothed in a brown, unspotted coat and weigh 5–9 kg (11–20 lb) with males slightly larger than females. The precocial calf performs a wobbly walk after 30 minutes and outruns most humans after 24 hours. Although other cervid mothers hide newborn calves and return periodically to nurse, a caribou calf continually accompanies its mother as she searches for food. Young calves suckle briefly and often—about twice per hour for less than one minute each time. A calf begins grazing by two weeks and continues to nurse for at least another four weeks and possibly into the winter. Age at sexual maturity is highly dependent on physical condition and ranges from 16 to 41 months.

Calf mortality is high and results from a combination of harsh weather and predation by wolf and lynx. Predation pressure on healthy adults is slight, yet continual, and comes in the form of the gray wolf. Hunting also is a major mortality factor for adults, and a few succumb to the meningeal worm. One caribou lived more than 20 years in a zoo, and maximum life span in the wild is about 13 years.

Suggested References: Darby and Pruitt 1984; Lavigueur and Barrette 1992; Miller 1982; Reimers 1993.

Vanished Species

Wolverine (*Gulo gulo*): A wolverine belongs to the Mustelidae and looks like a cross between a large weasel and a small bear. It has coarse, brown fur with two broad bands of lighter hairs extending across the shoulders and along the sides to the rump. Light-colored blotches often occur on the chest. The tail is long and bushy for a mustelid, and this mammal has large, heavily clawed feet. It generally weighs 11–16 kg (25–35 lb).

The wolverine roams the tundra and boreal forests of the Old and New World. Its original range may have included the entire Great Lakes region, but it was eliminated from most of this area prior to 1900. Today, its North American range covers northern Canada, Alaska, and a few other western states.

Home range varies from 100 to a staggering 2,000 km^2 (40–775 mi^2). A wolverine scent marks its territory and is highly aggressive toward others of the same sex. It often dens in a cave, among boulders, or in a simple depression underneath snow. It is active primarily at night and year-round.

The wolverine feeds mainly on carrion, including furbearers caught in traps, and occasionally raids remote cabins in search of food. The carnassial pair is extremely well developed, as is the temporalis, the main jaw-closing muscle, and these features give the wolverine a powerful bite, enabling it to feed on frozen carcasses and to crush large bones. Although not a swift runner, it sometimes attacks white-tailed deer or caribou, particularly during winter, when the large feet of the wolverine allow it to travel over snow with less effort than its prey. In the summer, it is more omnivorous and adds fruits, roots, birds, and bird eggs to its diet.

A wolverine mates from late April to July. Although gestation lasts 8–9

Wolverine. (Photo by Dallas A. Sutton.)

months, active growth of the embryo occurs only during the last 30–40 days. Most litters contain 2–4 young weighing 100 g (3.2 oz) each. Zoo animals live up to 17 years. Predation is rare, and only large bears and wolves are capable. Suggested References: van Zyll de Jong 1975; Wilson 1982.

Mountain Lion (*Felis concolor*): The mountain lion is the largest North American cat, with some mature males approaching 100 kg (220 lb). It has a yellowish brown coat, lighter fur on the belly, and a whitish chest and throat. Total length ranges from 1,500–2,700 mm (59–106 in), and the long, black-tipped tail measures 550–900 mm (22–35 in).

At one time, the mountain lion had the most extensive distribution of any mammal in the New World, ranging across North America and into South America as far as Argentina. Deforestation, overhunting of lion prey, and competition with humans for imported artiodactyls (cattle, sheep, etc.) contributed to the big cat's decline. Documented records of the mountain lion in the Great Lakes basin are scarce after 1850, although the species may have lingered in remote northern areas until the early twentieth century. Rumors still crop up concerning mountain lions roaming the dense woods surrounding Lake Superior, but none has ever been verified.

A mountain lion forages in rocky terrain or forested areas where white-tailed deer or other cervids are abundant. It is a solitary hunter that silently stalks unwary prey before leaping onto the victim's back and killing with powerful bites to the neck and shoulders. Usually, the lion eats part of the carcass and caches the rest under a layer of leaves and dirt. This cat is primarily nocturnal and has a home range of 30–300 km^2 (12–115 mi^2). In good habitat, the mountain lion covers 5–15 km (3–9 mi) each night in its search for food.

Although births occur in any month, they are most common in June and July. Females breed every other year and produce litters of 2–4 spotted kittens after a gestation of 92–96 days. Youngsters weigh 225–450 g (7–14 oz) at birth and breed during their second or third year. A mountain lion fears no animal except humans. Maximum longevity in captivity is 20 years. Suggested References: Currier 1983; Gerson 1988.

American Bison (*Bison bison*): Male bison weigh up to 900 kg (1,980 lb), making this species the largest land mammal in North America. Shaggy brown fur on the shoulders, front legs, and head give the animal a disheveled appearance. Its massive head, shoulder hump, and beard are unmistakable. It is the only species of Bovidae ever to occur here naturally, and like many in that family, both sexes grow short, curved horns.

The bison prefers grasslands or open woodlands; consequently, it was never abundant in the heavily forested Great Lakes region. Its original distribution skirted the southern shores of Lakes Ontario, Erie, and Michigan before turning northwest and continuing onto the prairies of middle America and Canada. European settlers found these large bovids easy targets, and they were extirpated from most eastern states before 1800. Today, bison survive in large preserves located in the West, in zoos, and on a few ranches, where they are raised for meat.

In the wild, females live with 5–20 other cows and their calves, whereas males are solitary or associate in small groups. During migration or at prime feeding sites, small bands associate to form herds numbering in the thousands. Bison are grazers, feeding primarily near dawn and

Mountain lion

dusk and spending the rest of the day chewing their cud. They cover only 3 km (5 mi) during daily wanderings but travel several hundred kilometers during the autumn and spring migrations. These bovids are able to swim across rivers that are 1 km (0.6 mi) in width and gallop at speeds up to 51 km/hour (32 mi/hour).

From July to September, males butt heads to establish dominance and the right to mate. Copulation takes place between April and June, and a single calf is dropped following 285 days in utero. The precocial offspring weighs 15–30 kg (33–66 lb) at birth, is weaned after 7–12 months, and becomes sexually mature when 2–4 years old. Wild individuals have lived up to 20 years. Suggested Reference: Meagher 1986.

American bison. (Photo by Gilbert L. Twiest.)

Capturing Small Mammals

Mammalogists most often work with small mammals, those that are the size of a mouse or smaller. Although there may be little glamour associated with mice and moles, there are many practical reasons for preferring small-bodied mammals. For one, they are easier to handle; a person may hold a bat or a shrew in one hand, but a wolf or bear requires considerably more effort. Second, small mammals generally are more abundant; a woodlot, for example, may harbor hundreds of shrews but only two or three deer. Third, biologists who work with small mammals seldom compete with fur trappers and hunters, and landowners typically are quite willing to have you take the little beasts off their hands. Finally, many studies require the use of captive animals, and it is much easier to feed and house 100 mice than it is to do the same for 100 moose.

Unfortunately, small mammals are secretive, often nocturnal, creatures. Most spend their lives within a jungle of leaves and grass, under rocks, in hollow logs, or in dark, subterranean tunnels. Although this life-style shelters small mammals from potential predators, such habits also hide them from curious biologists. Consequently, even just to discover what mammals live in a particular area requires setting traps. Trapping is often hard work. It is, however, a good excuse to get outdoors, and it is even a lot like Christmas—one never knows what will appear in the next package (trap).

Ethical, Legal, and Health Considerations

All mammalogists have a responsibility to their community, to themselves, and to the animals that they study to act in a professional manner. Every state and province has regulations that cover capturing, killing, and/or collecting wildlife, and all require that field biologists apply for a "scientific collecting permit" *before* beginning any field study. Regulations vary, and it is up to individual biologists to contact the appropriate agency (usually the Conservation Department or Natural Resources Department) for permission to work in areas under its jurisdiction.

Working with wildlife generally is a safe activity for the reasonably prudent person, but there are potential health hazards (Constantine 1988; Irvin and Cooper 1972). For example, Lyme disease is occasionally transmitted to humans through the bite of the deer tick, a parasite found on deer and small mammals, such as the white-footed mouse. Tularemia is a bacterial disease that may infect the eastern cottontail and may be passed on to a human who handles an infected carcass. Striped skunk and raccoon are two of the many species that occasionally suffer from outbreaks of rabies.

These are just examples; potential hazards vary with species, from region to region, and among years. It is *your* responsibility to contact a physician or the state, provincial, or local health department to determine the problems, risks, and necessary precautions in your area. In any event, always wear leather gloves to protect against animal bites and seek prompt medical attention if bitten. Discuss with your physician the advisability of immunizations against rabies (especially if working with carnivores or bats), tetanus, or other diseases. Always wear rubber or surgical gloves when handling dead animals and when preparing museum specimens.

[NOTE ADDED IN PROOF: Recently the U.S. Centers for Disease Conrol and Prevention (C.D.C.) attributed a cluster of human deaths in the southwestern United States to a hantavirus apparently contracted from rodents, most likely the deer mouse (Nichol et al., *Science* 262:914–17). This virus, or one similar to it, may be present in rodents of the Great Lakes region as well; however, the commonness of the virus, its virulence, and its potential to be passed on to humans are not known at the time that this book is going to press. Anyone contemplating work with rodents should contact a physician or the C.D.C. to learn the current status of this virus in the Great Lakes region *before* having contact with wild rodents.]

Finally, anyone working with mammals has a duty to treat these living creatures with respect and compassion. Always design your study to minimize their discomfort. The pamphlet "Acceptable Field Methods in Mammalogy," published as a supplement to the *Journal of Mammalogy*

(Choate 1987), contains many valuable suggestions for the humane treatment of mammals.

Live Traps

A variety of live traps, suitable for catching shrew- and mouse-sized mammals, are commercially available. Perhaps the most popular of these is the Sherman trap (fig. 21a). It is made from solid galvanized steel or aluminum and usually measures 8 x 9 x 23 cm (3.0 x 3.5 x 9.0 in), although smaller and larger sizes are available. The door of the trap folds in and down, compressing a spring, and is held in place by a small latch; the back of the trap contains a treadle that, when depressed, releases the door allowing it to snap shut. Small mammals are lured into the trap with some kind of bait, usually a mixture of peanut butter and oatmeal. These traps come as either a rigid box or in a form that folds for easy storage. One may obtain Sherman traps directly from the manufacturer (H. B. Sherman Traps, Inc., Tallahassee, Florida) or through scientific supply companies. Traps made from wire mesh (fig. 21b) are also available for mouse-sized mammals, but such traps most often come in sizes suitable for capturing larger animals, ranging from squirrels to raccoons. Wire-mesh traps also may be purchased from scientific supply companies and sometimes from local hardware stores.

For those on tight budgets, a modified Fitch trap (Fitch 1950; Rose 1973) is a homemade alternative. Figure 22 shows the principle behind this trap. The rectangular body is made from folded hardware cloth or other sturdy wire mesh; one end remains open for the entrance, and the other is attached with wire to a discarded can (#10) having a removable plastic cover. Rose (1973) suggests making the body from a square of hardware cloth (1-cm or 3/8-in mesh) that measures 30 × 30 cm (12 × 12 in). When folded, this produces an elongate trap almost 8 cm (3 in) on each side. Sheet metal for the door is bent inward at the sides, and a hole is drilled through the top of each side; the door is hung from the body with a horizontal length of wire that also acts as a hinge.

The trigger mechanism consists of two parts. The first of these is a piece of wire mesh that acts as a gate; it is suspended from the rear of the cage and is free to swing forward and back. The second part is a length of wire (18 gauge). At one end, this supports the door in an open position,

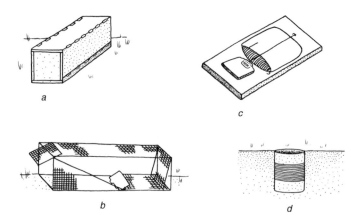

Fig. 21. Traps for capturing small terrestrial mammals: a) Sherman trap, b) mesh trap, c) snap trap, and d) pitfall trap

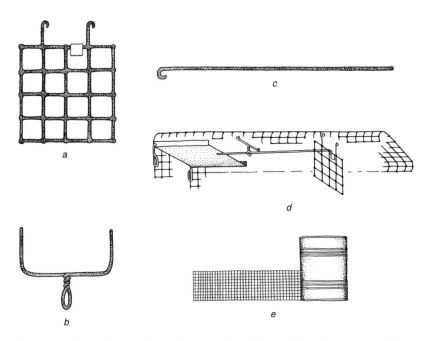

Fig. 22. Making a live trap for small mammals: a) the gate, b) a wire support, c) the wire that holds the door open, d) close-up showing parts in position holding door open, and e) distant view showing rectangular portion attached to tin can. (Modified from Rose 1973.)

and the other end loops over the gate. A small piece of sheet metal, firmly folded over the top of the gate, minimizes lateral movement of the trigger wire. The trigger wire also is supported by a stationary loop of wire placed 2.5 cm (1 in) behind the edge of the open door. Rose (1973) recommends constructing the trap so that the trigger wire and the open door are parallel to the floor. If bait is placed in the rear of the cage, an interested animal pushes the gate on its way to the bait, thus pulling the trigger wire slightly inward and releasing the door.

A Fitch trap captures many types of mice and medium-to-large shrews. Do-it-yourself instructions also exist for making specialized traps for ground squirrels (Shemanchuk and Bergen 1968) and moles (Baker and Williams 1972; Yates and Schmidly 1975). In addition, Taber and Cowan (1971) suggest a number of trap designs for large mammals.

Snap Traps

Mammalogists frequently must collect specimens; these are essential for studying anatomy, geographic variation, taxonomy, food habits, internal parasites, and a host of other biological parameters. The most common, and one of the most humane, ways of obtaining specimens is by snap trapping. These traps (fig. 21c) often catch species that rarely enter or trigger live traps; hence, they are also useful in documenting which species live in a particular area. Snap traps are familiar to most people, easy to use, quite effective, and much less expensive than live traps. One Sherman trap, for example, costs 3–20 times more than a single snap trap.

Snap traps come in three sizes. Mammalogists use "rat" traps, 17 cm long × 8 cm wide (7 × 3 in), to capture rats and most squirrels. The "mouse" trap, in contrast, is much smaller, 10.0 × 4.5 cm (4.0 × 1.8 in), and readily takes shrews, mice, jumping mice, and voles. However, mouse traps frequently crush the skull, which is one of the most valuable parts of the specimen; consequently, many collectors prefer an intermediate-sized trap called the Museum Special. This trap kills by breaking the animal's back, rather than crushing the skull, and readily captures mammals weighing up to 50 g (2 oz). Peanut butter and oatmeal are standard bait when using snap traps, and biologists usually combine the two until the mixture is dry enough to handle easily but still moist enough to stick to the trap. Hardware stores commonly stock rat and mouse traps, and Museum Specials may be purchased from Woodstream Corporation (Lititz, Pennsylvania).

Sampling Small Mammal Populations with Live Traps or Snap Traps

Biologists sample the small mammals living in a particular habitat by establishing a trap line. While walking in a straight line, set traps either at regular intervals, such as every 10 m (30 ft), or place traps at likely locations. Traps set near burrow entrances, adjacent to logs, in hollow stumps, or along stream edges are particularly productive. In meadows and prairies, gently part the grass until you find the tunnellike passageways built by voles and used by many other species; pieces of freshly cut grass or feces indicate current use. Place traps perpendicular to the runway so that a mouse coming from either direction passes by the live-trap opening or over the snap-trap trigger. Be sure that overhanging vegetation does not interfere with the action of the trap. Flying squirrels and some of the more arboreal mice may be captured by placing a trap on a partially fallen tree inclined at about a 45° angle.

It is very easy to lose traps. Therefore, mark the location of each with a piece of cloth or flagging tape conspicuously tied to a high weed or nearby tree. In addition, consecutively number each trap station so that you can keep track of the traps. A forgotten trap means lost money but, more importantly, means that an animal will die for no useful purpose. An opossum, raccoon, or other scavenger occasionally walks away with the trapped mammal, as well as the trap, so many collectors secure the trap to a root, branch, or stake with a length of stout string or wire.

Most mammalogists set traps at dusk and retrieve their catch the following morning. If traps are set during the day, check them periodically. Specimens caught in snap traps spoil rapidly in the sun, and day-active invertebrates, such as ants, frequently devour the bait or damage the mammal carcass. Live traps exposed to direct sunlight overheat and kill the animal inside, so shade these traps with a board or leafy branch. When live-trapping during cold weather, add extra food (bait) to live traps, as well as cotton batting or other nesting material, to insure the animal's survival.

Pitfall Traps

Pitfall traps are nothing more than metal cans or plastic containers, about 15–20 cm (6–8 in) deep and 9 cm (3.5 in) or more in diameter, buried so that the rim of the can is flush with the ground (fig. 21d). Small mammals wandering by simply tumble into the trap and cannot escape because of the steep, smooth walls. Trap success is higher if animals are guided toward a pitfall by a "drift fence," a long strip of aluminum flashing or plastic sheeting set vertically into the ground; a rodent or shrew that encounters the flashing generally turns and follows along it until the mammal drops into the trap. If a drift fence is not available, a pitfall placed next to a log or large rock is almost as good. A long-handled bulb planter, available from garden-supply stores, provides a quick and easy way of digging the necessary holes for small pitfall traps. If using pitfalls as live traps, provide plenty of food and check them frequently—if two animals fall into the same trap, one may devour the other. Some collectors turn pitfalls into kill traps by adding 5–10 cm (2–4 in) of water. Although pitfalls require more labor than setting a Sherman trap or snap trap, they are very effective, especially at capturing small shrews (Williams and Braun 1983).

Harp Traps and Mist Nets for Bats

A harp trap (fig. 23a) is an excellent way of capturing bats as they enter a barn or cave (Kunz and Kurta 1988). This trap consists of two aluminum frames, each about 1.5 m × 1.8 m (5 ft × 6 ft), and separated from each other by 7–10 cm (3–4 in). Along each frame, about 2.5 cm (1 in) apart, are strung vertical lengths of six-pound monofilament fishing line. Bats attempting to fly through the trap make it through the first set of lines, but not the second, and ultimately fall unharmed into a canvas bag hung from the frames. Tuttle (1974) and Tidemann and Woodside (1978) provide directions for building such a trap.

Although biologists developed the harp trap specifically for capturing bats, mist nets originally were designed for catching birds. These nets are made from thin nylon, 5.5–18.0 m (18–60 ft) wide, and typically only 2 m (7 ft) high, but biologists often stack nets on top of one another to increase the effective height. For capturing bats, mist nets work best when used along a corridor, such as a stream with wooded banks or a

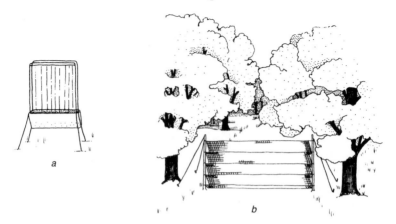

Fig. 23. Harp trap (a) and mist net (b) for capturing bats

roadway through a forest (fig. 23b). Stretch the net from the trees on one side, across the road or creek, and into trees on the other side. Place the bottom of the net near the ground or water and extend the top as close as possible to overhanging branches. Such placement effectively blocks the available flight space and greatly increases your chances of capturing a bat.

Most bats are active during the first 1–4 hours after sunset, and netting is most productive during this time. It is a waste of time to net if air temperature is below 10°C (50°F) or if it is raining; bats are sensible creatures and usually do not fly under such conditions. Check nets at least every 20 minutes to prevent undue stress on the bats and minimize holes chewed into your net. Take extreme caution when extracting a bat from a mist net because a bat's arm and finger bones are very delicate and easily snapped by inexperienced workers. Remember to wear leather gloves, and be sure to read Constantine's (1988) article on health precautions before handling bats.

Currently, only the Japanese manufacture mist nets, and their availability is tightly controlled by the Japanese government. Contact your local Audubon Society chapter or consult Kunz and Kurta 1988 for possible sources of mist nets; you must possess a scientific collecting permit in order to purchase nets. Kunz and Kurta (1988) also provide many tips on the art of mist netting, as well as other techniques used to capture bats.

Specimen Preparation

Field biologists often collect specimens that died from natural causes or a sudden encounter with a fast-moving automobile or snap trap. Mammalogists carefully preserve such specimens in the field and eventually incorporate them into a research collection at a museum or a teaching collection at a high school or college. Below are some suggestions for measuring, labelling, and preparing small mammal specimens. For information on preserving large mammals or bats, consult Nagorsen and Peterson 1980 or Handley 1988, respectively. Be sure to consider potential health hazards (p. 280) before handling wild mammals.

Measuring Specimens and Recording Data

To a professional mammalogist, data accompanying a specimen are possibly more valuable than the animal itself. Record the following information on a sturdy label using a hard pencil or waterproof ink (fig. 24).

1. Date of capture. Avoid ambiguity by writing out the month and the complete year (e.g., 6 September 1994, not 9-6-94 or 6-9-94).
2. Capture locality. Include state or province, county, and the airline distance and direction to the nearest town or other obvious landmark.
3. Collector's name.
4. Specimen number. Mammalogists sequentially assign a number, beginning with 1, to each specimen that they prepare and never duplicate numbers, not even between years.
5. Sex of specimen. Use ♂ for male and ♀ for female.
6. Reproductive condition. Indicate whether a female was pregnant or lactating; if pregnant, record the number of embryos and their size.

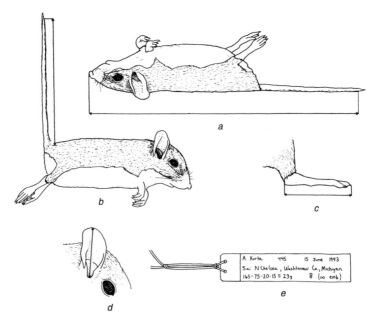

Fig. 24. Standard measurements and a typical label: a) total length, b) tail length, c) hindfoot length, d) ear height, and e) a label

For males, indicate the size and location of testes. Nagorsen and Peterson 1980 or Taber 1971 provide tips on determining reproductive condition.

7. Standard measurements (see fig. 24). List linear measurements in order, each separated by a dash, then draw three horizontal lines followed by the mass.

   a. Total length. Lay the animal flat without stretching it, and measure the distance from the tip of the snout to the end of the tail vertebrae. Do not include any hairs that extend beyond the last tail vertebra.

   b. Tail length. Bend the tail up at a 90° angle, and determine the distance from the tail base to the end of the tail vertebrae.

   c. Hindfoot length. Measure this from the heel to the tip of the claw on the longest toe.

   d. Ear height. This is the distance from the lowest notch at the base of the ear to the highest point of the ear pinna, excluding hairs.

   e. Body mass. Weigh the specimen if possible.

Mammalogists also enter the same information into a notebook, or catalog, thereby keeping a permanent record of their lifetime catch.

Preparing a Study Skin

Taxidermic mounts, although visually pleasing, take too much time to prepare and require too much storage space. Instead, mammalogists usually prepare animals smaller than a woodchuck as a "study skin," retaining the outer shell of the animal and stuffing it with a cotton form. These instructions should start you on your way, but remember that making a good study skin requires patience and considerable practice. Before beginning, thoroughly read the instructions, and be sure to have all necessary equipment and supplies within easy reach. Most field biologists pack the items listed here, along with materials needed for measuring specimens and recording data, into a kit made from a fishing tackle box. To make a study skin, you need:

1. dissecting tools (forceps, scalpel, scissors) to separate the skin from the body,
2. needle and white thread for sewing the final specimen,
3. wirecutters and wire (preferably noncorrosive) to support the tail and legs,
4. cotton (preferably nonabsorbent) for stuffing the specimen,
5. cornmeal or hardwood sawdust to absorb grease and body fluids,
6. pins and cardboard or soft wood for positioning the specimen as it dries,
7. an old toothbrush for smoothing and cleaning the fur, and
8. rubber or surgical gloves to protect your hands.

You should prepare study skins as soon as possible after the animal's death to prevent the fur from falling out in patches, or "slipping." You can freeze freshly killed mammals for later preparation, but wrap them tightly to minimize desiccation.

Removing the Skin

1. After donning gloves, make an incision through the skin in the midabdominal area; pinch the skin with forceps or fingers, pull up gently, and

make a small initial cut with scissors. Extend the cut from near the anus to the sternum (breast bone), generally working along the midline, but passing to the side of the penis or vagina. Avoid cutting the underlying muscles, and, if working with a mustelid, be careful not to puncture the anal scent glands.

2. On either side of the incision, gently work the skin loose from the body wall and hindlegs, using a blunt probe or gloved finger. Continually sprinkle liberal amounts of cornmeal on the underside of the skin and any exposed body parts; this absorbs grease, blood, and other body fluids and makes the skin easier to grip. While holding the hindfoot, force the knee out the incision and sever the leg at the hip or knee joint (fig. 25a). Trim all remaining muscle down to the heel; if left on the bone, the meat likely will rot before it dries. Do the same for the opposite leg, and then snip the connections to the digestive and urogenital tracts. Don't forget to use cornmeal.

3. Now you are ready to remove the tail vertebrae. Firmly grasp the base of the tail with stout forceps or the nails of the thumb and one or two other gloved fingers (fig. 25b), and pull the vertebrae out with the other hand. The vertebrae should slip out while the tail skin bunches behind the fingers; do not allow the tail skin to turn inside out. If the vertebrae do not yield under slight pressure, loosen the connective tissue by firmly rolling the tail on a flat surface. For squirrels and larger animals, you probably will have to remove the vertebrae through a longitudinal incision on the ventral side of the tail and later sew the tail skin back together.

4. After severing the tail and legs, start peeling the skin from the body (fig. 25c). Gently loosen the skin and avoid stretching it. Many beginners tug so strongly that the final product looks like a weasel, regardless of the animal's original shape. When you reach the forelegs, sever them at the shoulder or elbow and trim the muscle as you did the hindlegs. Continue rolling the skin toward the head, and keep adding cornmeal.

5. When you reach the ears, identify the cartilaginous bases, and sever each ear canal as close to the skull as possible (fig. 26a). Carefully work forward until the dark eyes are visible through a thin tissue, and cut through this membrane (fig. 26b). The skin will remain attached at the inside corner of the eye. Cut this attachment as well, and avoid nicking the eyelid itself; if you do, the skin may rip when the body is finally stuffed. Next, peel the skin toward the lips, and cut the connective tissue that attaches them to the skull. The only remaining attachment is at the

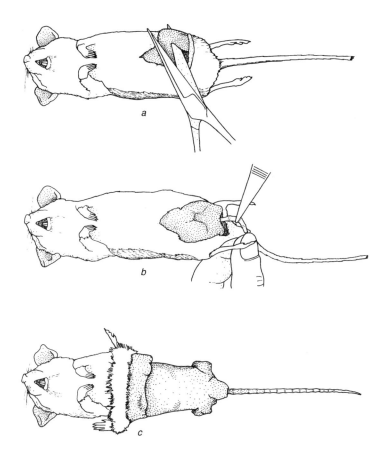

Fig. 25. Early stages in skinning a small mammal: a) severing leg bones, b) removing tail vertebrae, and c) rolling back skin

nasal cartilage. Cut this cartilage (fig. 26c), taking care not to damage the nasal bones, and finally separate the skin from the carcass. Place the carcass aside for now and finish preparing the study skin.

6. Lay the skin flat and pick off any fat, glandular tissue, or remaining muscle. Some mammalogists dust the skin with various chemicals that hasten drying and provide some protection against future insect damage, but this is not a necessity (see Nagorsen and Peterson 1980 for details). After cleaning the skin, sew the lips shut with a triangular stitch, as shown in figure 26d. If the skin begins to dry out before you finish, slightly

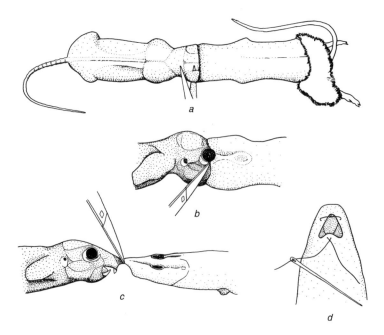

Fig. 26. Later stages in skinning a small mammal: a) separating ear from skull, b) cutting around eye, c) separating skin from nasal cartilage, and d) a triangular stitch

moisten the flesh side with a water-soaked cotton swab. If you noticed any blood stains on the fur, now is a good time to remove them; use a cotton swab dampened with water or alcohol, and soak up excess solvent with cornmeal. The skin should be inside out to begin stuffing.

Replacing the Body with a Cotton Form

7. To form the body, take a single piece of cotton and loosely roll it into a cylinder that is slightly greater in diameter and longer than the skinned mammal. To form a narrow snout, press down on the cylinder in the center of one end with forceps, fold the two corners over to the midline, and regrasp with forceps (fig. 27a). While maintaining your grip, place the pointed end of the cylinder next to the nose pad of the still everted skin and slowly work the skin back until it is right side out and surrounds the cotton form (fig. 27b). Do not release the point until you are satisfied that it extends all the way forward and that the skin of the head is

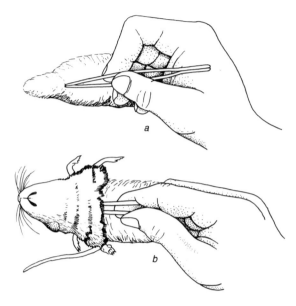

a

b

Fig. 27. Filling the skin with a cotton form: a) roll of cotton with head formed and ready to be inserted, and b) placing cotton form inside and turning skin right side out

symmetrical and snug around the cotton. Now continue until the entire skin is stuffed and totally fur side out.

8. You should support each leg with a piece of straight wire that is long enough to extend from the foot to midbody. Gently insert the wire under the skin, along the sole, and as far as the longest toe. Wrap the wire and leg bones together with cotton; this secures the wire in place and fills in the area where muscles were removed. If the animal is mouse sized or smaller, you may simply add wire without cotton. (Some preparators prefer to add leg wires before stuffing the body.)

The diameter of wire needed varies with the size of the mammal. Try 26 gauge for the smallest shrews and bats, 22–24 gauge for large shrews and most mice, and 18–20 gauge for small squirrels. You can straighten wire from a spool by securely tying one end to an immoveable object, grasping the other end with pliers, and then pulling.

9. Once leg wires and the cotton body are in place, you need to fashion a support for the tail. Cut a piece of wire long enough to extend from tail tip to midbody. Pull out a thin wisp of cotton, moisten one end of the wire, and place a few cotton strands around it near the end (fig. 28a); the moisture helps keep the cotton in place. Next twirl the wire

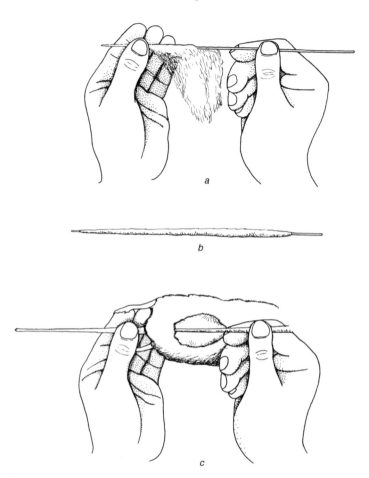

Fig. 28. Making and inserting a wire support for the tail: a) placing cotton on wire, b) completed tail wire with tapered shape, and c) inserting tail wire

with one hand and gradually feed more cotton onto the wire with the other. Continue to add strands until the wrapped part matches the general shape of the animal's tail in life (fig. 28b). Take the wrapped wire, moisten it slightly, and gently insert it into the empty tail sheath (fig. 28c). If it sticks on the way in, you will have to rewrap the wire with less cotton. When you are finished, the free end of the wire should lay along the ventral body surface, between the cotton body and the skin. Wrapping a tail wire often is a frustrating experience for beginners, and you should practice before actually working on your first specimen.

Finishing Touches

10. Make sure that the legs and tail are straight and parallel to the midline and then sew the belly incision starting at the anterior end (fig. 29a). Use a series of loose diagonal stitches, about 10–15 mm (0.25–0.50 in) apart, alternating from right to left side. When you reach the end of the incision, snug the stitches until the sides just meet and securely knot the thread before cutting it. Now take the label that you prepared while measuring the animal, and tie it to the right hindleg of the specimen, just above the ankle.

11. Remove most cornmeal or other debris adhering to the skin by blowing on it or gently brushing with an old toothbrush. Using the toothbrush, smooth the ventral fur so that the hairs lie in a natural position.

12. Lay the specimen belly down on a piece of corrugated cardboard or soft, porous wood, and place the forefeet sole down, next to or slightly under the head, with the limbs parallel to the body axis (fig. 29b). Attach the specimen to the cardboard by inserting a pin through each foot; be sure that the feet are symmetrical, that is, the same distance forward and the same distance from the side of the head. Next, straighten the body and tail, and secure the tail tip by crossing two pins over it in the form of an **X** (fig. 29b); an additional pair of pins may be used to position the base of the tail. Be sure that the pinned tail is aligned with the midline of the body, and then pin the hindfeet sole down, next to and essentially parallel with the tail. Use forceps to pull a small bit of cotton through the eyes so that they do not dry shut, and brush the animal's back and sides with the toothbrush. Check the symmetry of the pinned specimen one last time, and make any necessary corrections, because once the specimen dries, no changes are possible.

13. Place the pinned mammal in an area of good ventilation, out of direct sunlight, and protected from insects. Depending on humidity and the size of the animal, the specimen will dry within one to seven days. When the skin is dry, remove the pins by twirling them and place the specimen in an insect-proof container or donate it to a museum.

Saving the Skull

14. Most biologists save the animal's head, later cleaning the skull and incorporating it into a collection along with the skin. Sever the head from the neck vertebrae being careful not to damage the back of the skull, and

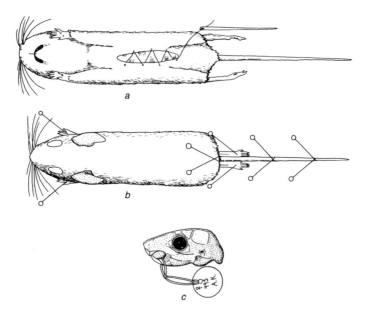

Fig. 29. Final steps in preparing a skin and skull: a) sewing incision, b) pinning specimen, and c) a skull label

tie a label around one side of the lower jaw. A suitable skull label contains the collector's name or initials, the catalog number that corresponds to the skin of the specimen, and usually, the sex of the animal (fig. 29c).

15. A skull must dry quickly or else it begins to rot, resulting in objectionable odors and possibly making it difficult to clean later. To hasten drying of a squirrel-sized skull or larger, you should cut out the tongue and eyeballs, and trim much of the jaw musculature. Also remove the brain using a hooked piece of wire inserted through the foramen magnum, or flush out this tissue with water from a blunt syringe. A small skull, such as that from a bat or mouse, generally dries fast enough without this extra attention. Place the skull in an area with good air movement and protect it from flies with cheesecloth or window screening; if fly eggs successfully hatch, maggot activity eventually separates the bones, severely lessening the value of the specimen. Small skulls generally dry within a few days and can be kept for weeks or months before final cleaning. Simple instructions for cleaning a skull are given later on.

16. If you are saving the entire skeleton, remove all internal organs and trim muscles of larger specimens as well. Be sure to attach an additional label to the body.

Fluid Preservation

Preserving small mammals in fluid is simple and makes available all parts of the animal for future reference. The first step involves "fixing" the specimen; a good fixative not only prevents bacterial degradation but also halts the chemical breakdown processes of individual cells. The most popular fixative is formalin, which is commercially available as a 37–40% solution of formaldehyde gas dissolved in water. For preserving animals, use a 10% solution of formalin (9 parts water and 1 part formalin). (**CAUTION**: whenever using formalin, wear skin and eye protection, work with adequate ventilation under a fume hood, and follow the manufacturer's precautions.)

Formalin is acidic and eventually decalcifies bones and teeth; consequently, many collectors add chemicals to neutralize the formalin (Jones and Owen 1987). If available, add 4 g of monobasic sodium phosphate monohydrate ($NaH_2PO_4 \cdot H_2O$) and 6.5 g of dibasic sodium phosphate anhydrate (Na_2HPO_4) to each liter of 10% formalin; this will maintain the pH near 7.0 for short periods.

Small animals should be immersed in fixative as quickly after death as possible. Use a container large enough so that solution completely covers the specimen and an amount of fixative that is 6–10 times the volume of the animal. If you use less, water from the mammal's tissues rapidly dilutes the formalin, and decomposition may occur. After measuring the specimen, attach a waterproof label. Prop the mouth open with a piece of wood or other nonmetallic object; if not done, jaw muscles harden in a closed position, making it impossible to examine the teeth without damaging the specimen. Next, make a ventral incision through the abdominal skin and underlying muscles in order to speed contact between the fixative and internal organs; palpate the abdomen after immersion to remove trapped air and facilitate the entry of formalin. As an alternative to slitting the belly, use a syringe to inject solution into the body cavity until the abdomen is firm and then immerse. (**CAUTION**: improper use of a syringe may cause its contents to be sprayed on the user.) After 1–4 days in fixative, the specimens are hard to the touch and should be transferred to ethyl alcohol (70–75%) for long-term storage (Jones and Owen 1987). Fluid preservation of mammals larger than a chipmunk requires perfusion, the injection of fixative directly into the circulatory system; consult Forman and Phillips 1988 for suggestions.

Alternative Techniques for Teaching Collections

Wet/Dry Preservation Using Formalin and Borax

This is an alternative preservation technique that is suitable for some types of small specimens that are destined for a school collection (Sheridan 1978). First take formalin and add borax (sodium tetraborate) until the solution is saturated. (**CAUTION**: whenever using formalin, wear skin and eye protection, work with adequate ventilation under a fume hood, and follow the manufacturer's precautions.) Using a syringe with an appropriately sized needle, inject the solution into the abdominal cavity, thoracic cavity, and brain; major muscles and even the tail may need to be injected on specimens larger than a mouse. (**CAUTION**: improper use of a syringe may cause its contents to be sprayed on the user.) Make multiple injections to insure that all tissues are saturated and allow the solution to diffuse into the tissues before proceeding. Avoid using too much of the solution; for example, only 1–2 ml is sufficient for the abdominal cavity of a deer mouse or short-tailed shrew. Using larger amounts or moving the carcass immediately after injection invariably results in fluid leaking from the injection site and borax eventually precipitating on the fur. After the solution is absorbed, pin the animal as you would a study skin (see fig. 29) and allow it to dry in an area with good ventilation. The formalin eventually evaporates, leaving behind the borax to act as a preservative. Although quicker than making a study skin, there are drawbacks to this technique; the skull remains in the skin, and borax may affect fur color, especially reds. Also, tissue surrounding the abdominal cavity may collapse slightly during the drying process, resulting in a specimen less visually pleasing than a study skin.

Do-It-Yourself Tanning of Small Mammal Skins

Most museums send skins to commercial tanners, but the following technique provides reasonable results with animals such as muskrat, mink, and fox. Begin by skinning the animal and clearing away any fat, muscle, or other tissue that adheres to the underside of the skin. Vigorously rub salt (NaCl) over the inside surface of the skin to help dry and temporarily preserve the skin; it may be left with the salt on, for days even, until you are ready to go on. Next, dissolve equal amounts of salt and alum in water until you have a saturated solution, and then soak the skin for about one week. When ready to proceed, mix together 6–8 egg yolks and a small amount of flour with some of the salt-alum solution until you have a workable paste. Spread the paste thinly over the inner skin, and let it sit for a week or two before peeling off the paste. Now, work the skin back and forth over a table edge, fur side down, until it is soft and pliable. Excess oils may be removed with cornmeal or by washing in a mild detergent. If possible, minimize contact between fur and the salt-alum solution, because these chemicals may alter fur color.

Cleaning Skulls

Dermestid Beetles

Although most museums have a "bug room" for cleaning skulls and other skeletal material, the insects involved are really dermestid beetles and not true bugs. Adults are less than 12 mm (0.5 in) long and have shiny brown or black backs, whereas the brownish larvae are segmented and very hairy. In the wild, these beetles feed on dead animals, and in captivity, they eagerly remove flesh from bones provided by mammalogists. A large, well-maintained colony can clean a mouse skull in a matter of hours.

To start your own colony, gather dermestids from under any mummylike carcass in nature or purchase the beetles by mail from a biological supply company. Be sure to obtain both larvae and adults; although adults do very little cleaning, they are needed for reproduction. House the beetles in a large glass jar or hard plastic jug. A layer of cotton placed at the bottom of the container provides sites for larvae to pupate (Sommer and Anderson 1974). Cover the container with a screened lid to provide air flow and prevent escape of even the smallest larva. Never keep a colony in your home or near a collection, because escapees cause severe damage to household goods and animal specimens.

Skulls given to dermestids should be fairly dry. Bacteria readily decompose the flesh on wet skulls, producing objectionable odors, and any mold that grows may repel the beetles. Check the progress of the dermestids at least once each day, and remove the skull when most tissue has disappeared. You may have to trim the last few bits of muscle or ligament with scissors, scalpel, nail clippers, or forceps. Soaking the skull in a mild solution (5%) of household ammonia for a few hours may make this last step easier, as well as remove any grease; be sure to thoroughly rinse the skull with lukewarm water to remove all traces of ammonia. When the cleaning job is satisfactory, place the skull in a freezer for 3–7 days to kill any small dermestid larvae that you may have overlooked.

Boiling

To clean a skull with the aid of boiling, start with either fresh or dried material. Place the skull in a boiling vessel with enough water to cover the skull completely; this prevents any fat that floats to the top from accumulating on the specimen. Bring the water to a boil and simmer until the muscles appear gelatinous and begin loosening from the bone. Rapid cooling generally causes teeth to crack, so allow the pot to approach room temperature before pouring off floating grease and removing the skull. Most tissue should pull away easily, and the rest can be taken off with scissors or a scalpel. Boiling works well with squirrel-sized or larger specimens, but delicate skulls, such as those of shrews and bats, are often damaged.

Bacterial Maceration

Maceration is a euphemism for bacterial rot. It is a time-honored technique that, although slow and possibly odorous, produces excellent results. Place the skull in a glass container, cover with water, and store in a warm room for 1–4 weeks. Shake the jar every day or so, and completely change the water at least every week. During this time, bacteria thrive in the culture and decompose any soft tissue surrounding the skull. Check its condition periodically, and remove the skull when muscles and tendons can be picked away easily. As always, wear gloves when removing the flesh to prevent possible infections.

Degreasing and Bleaching

There usually is little grease associated with small mammal skulls, and what little there is often is removed by the earlier soaking in a 5% solution of household ammonia. If you find further degreasing necessary, consult DeBlase and Martin 1981 or Knudsen 1972 for suggestions. Be forewarned, however, most degreasing chemicals are highly toxic.

Although a cleaned skull destined for a research collection is never bleached, you may want to whiten specimens intended for display. To do this, simply soak the skull in a fresh solution of 3% hydrogen peroxide (available at any pharmacy) until attaining the desired whiteness. When finished, rinse the specimen in running water to remove all traces of the peroxide, and allow it to dry in a well-ventilated area.

Identifying Adult Mammals of the Great Lakes Region

Although one could identify a mammal by comparing it to the photographs and descriptions accompanying each species account, such an approach is time consuming; consequently, biologists frequently use a "key" as a simple alternative. A key offers two contrasting sets of characteristics at each step. By choosing the most appropriate set, the reader eventually is led, in a step-by-step process, to the correct identification of a specimen.

Below are two keys—one based on mammals in the flesh and one for skulls. Each of these, in turn, consists of two parts—a key to mammalian orders and a key to species within each order. The key to animals in the flesh only covers the 80 species that currently live in the wild, in and near the Great Lakes region. The key to skulls, however, also includes common domesticated mammals, as well as the three native species that have been exterminated.

By using the keys, you can identify most specimens using nothing more complex than a ruler; however, identification of a few small skulls (especially *Sorex*, *Myotis*, and *Microtus*) requires calipers for more exact measurements and a hand lens for examining teeth. Be sure to consult the glossary for definitions of technical terms, and use the numerous drawings to interpret unfamiliar characters. Examine all traits that are listed, not just the first, and use the "weight of the evidence" to guide your decision. Whenever possible, key out both the animal and its skull. The Indiana bat and little brown bat, for example, are very difficult to tell apart using skull characteristics but simple to identify based on external traits. The masked shrew, pygmy shrew, and southeastern shrew, in contrast, may be impossible to separate based on external features, but a quick look at the teeth usually clinches the identification. Also, referring to the geographic distribution of each species often eases the identification problem. For example, the northern and southern bog lemmings are very similar in external appearance; however, the northern bog lemming is found only near the northwestern boundary of the Great Lakes basin, and, anywhere else, one is reasonably confident that the specimen must be the southern species.

The keys in this book are an attempt to simplify the immense variation seen in nature and, of course, are not infallible. Immature animals, individuals that are much larger or smaller than average, color variants, damaged specimens, hybrids, and escaped exotic pets are particular problems. For identification of difficult specimens, consult a mammalogist at a local college or contact a major research-oriented museum. Within the Great Lakes region, the largest mammal collections are located at the Royal Ontario Museum (Toronto), Michigan State University Museum (East Lansing), University of Michigan Museum of Zoology (Ann Arbor), and Field Museum of Natural History (Chicago). The curator of mammals at each institution generally enjoys the challenge of identifying odd specimens.

Key to Adult Mammals Using Skull Characteristics

Key to Orders

1a. More than three incisors on each side of upper jaw; more than three incisors on each side of lower jaw; jugal bone helps form anterior part of jaw joint; angular process of lower jaw points inward (is inflected); greatest length of skull usually between 100 and 130 mm ....
.... **New World Opossums: Order Didelphimorphia, I.**
1b. Three or fewer incisors on each side of upper jaw; three or fewer incisors on each side of lower jaw; jugal bone not involved in jaw joint; angular process of lower jaw not inflected; greatest length of skull variable.............. 2.
2a (1b). Upper incisors not present ....
.... **Even-toed Ungulates, in part (Deer, Cattle, Bison):**
Order Artiodactyla, VII.

2b. Upper incisors present . . . . 3.

3a (2b). Skull large, greatest length usually much greater than 300 mm; eye socket (orbit) surrounded by complete bony ring that is circular in shape . . . .
. . . . **Horses: Order Perissodactyla, VIII.**

3b. Skull smaller, greatest length of skull less than 300 mm; orbit not surrounded by complete, circular, bony ring. . . . . 4.

4a (3b). Large toothless gap (diastema) between incisors and cheek teeth, especially evident on upper jaw; length of upper diastema greater than height of first upper incisor; first incisor enlarged, height greater than any other tooth, curving backward, chisellike; no canines present . . . . 5.

4b. Large diastema between incisors and other teeth lacking; incisors variable in size and shape; canines present, although not always well developed . . . . . 6.

5a (4a). Two upper incisors on each side, each lacking yellow or orange pigmentation on anterior surface; the second, small, peglike, located directly behind first; skull with many small openings (fenestrations), especially anterior to orbit. . . . . **Rabbits and Hares: Order Lagomorpha, IV.**

5b. One upper incisor on each side, usually (but not always) with yellow or orange pigmentation on anterior surface; skull not fenestrated . . . .
. . . . **Rodents: Order Rodentia, V.**

6a (4b). When viewed from below, right and left upper incisors distinctly separated by U-shaped notch in palate . . . . **Bats: Order Chiroptera, III.**

6b. When viewed from below, right and left upper incisors not separated by U-shaped notch in palate. . . . . 7.

7a (6b). Many teeth with at least some reddish brown (chestnut) pigmentation; zygomatic arch never present. . . . .
. . . . **Insectivores, in part (Shrews): Order Insectivora, II.**

7b. Teeth without chestnut brown pigmentation; zygomatic arch present . . . . . 8.

8a (7b). Chewing (occlusal) surface of rear cheek teeth with W-shaped ridges (top of W faces laterally); canine poorly developed (in these mammals, the upper canine is the *fourth* tooth from the front, whatever its size or shape); first upper incisor usually about equal in size or larger than canine . . . .
. . . . **Insectivores, in part (Moles): Order Insectivora, II.**

8b. Rear cheek teeth without W-shaped ridges; canine usually well developed (in these mammals, the upper canine is also the fourth tooth from the front); first upper incisor about 50% of the height of canine or smaller . . . . 9.

9a (8b). Roof of skull in back of orbits slopes upward to posterior margin; seven upper cheek teeth; canines triangular in cross-section; greatest length of skull usually much more than 175 mm . . . .
. . . . **Even-toed Ungulates, in part (Pig): Order Artiodactyla, VII.**

9b. Roof of skull in back of orbits slopes downward to posterior margin or remains at about same height; six or fewer upper cheek teeth; canines

rounded, not triangular in cross section; skull size variable . . . .
. . . . **Carnivores: Order Carnivora, VI.**

Key to Species by Order

I. Order Didelphimorphia: Family Didelphidae, New World Opossums . . . .
. . . . Virginia Opossum, *Didelphis virginiana*, p. 19, fig. 3.

II. Order Insectivora: Shrews and Moles.
1a. Teeth without chestnut brown pigmentation; slender zygomatic arch present
   (although often broken); greatest length of skull more than 30 mm. . . . .
   . . . . **Family Talpidae: Moles** . . . 2.
1b. Teeth with chestnut brown pigmentation; zygomatic arch never present;
   greatest length of skull less than 30 mm . . . . . **Family Soricidae: Shrews** . . . 4.
2a (1a). First upper incisor points down and forward; first lower incisor points
   forward, essentially horizontal; palate ends opposite or in front of last upper
   cheek tooth . . . . Star-nosed Mole, *Condylura cristata*, p. 56.
2b. First upper incisor essentially vertical; first lower incisor not horizontal; palate
   ends just in back of last upper cheek tooth . . . . 3.
3a (2b). Auditory bulla totally closed (complete); eight teeth in each lower jaw;
   greatest width of skull more than 16 mm . . . .
   . . . . Eastern Mole, *Scalopus aquaticus*, p. 54, fig. 5.
3b. Auditory bulla partly open on bottom (incomplete); 11 teeth in each lower
   jaw; greatest width of skull less than 16 mm . . . .
   . . . . Hairy-tailed Mole, *Parascalops breweri*, p. 51.
4a (1b). Three unicuspids (simple teeth between the large incisors and more
   complex cheek teeth) *readily* visible in lateral view of upper jaw (others present but small and totally or partly hidden) (fig. 30a–b) . . . . 5.
4b. More than three unicuspids *readily* visible in lateral view of upper jaw . . . 6.
5a (4a). First upper incisor without medial tine (fig. 31a, top); in ventral view,
   four unicuspids evident, the fourth minute and hidden from lateral view (fig.
   31b, top). . . . . Least Shrew, *Cryptotis parva*, p. 43.
5b. First upper incisor with well-developed medial tine (fig. 31a, bottom); in
   ventral view, five unicuspids evident, the third and fifth minute and often
   hidden from lateral view (fig. 31b, bottom) . . . .
   . . . . Pygmy Shrew, *Sorex hoyi*, p. 36.
6a (4b). Opening at front lateral side of mandible (mental foramen) located under
   posterior cusp (hypoconulid) of fourth lower tooth (fig. 31c, top); base of first
   lower incisor located behind anterior margin of fourth lower tooth (fig. 31c,
   top); greatest width of skull more than 11 mm. . . . .
   . . . . Northern Short-tailed Shrew, *Blarina brevicauda*, p. 46, fig. 4.

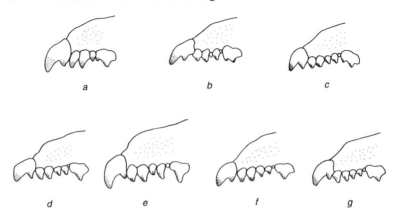

Fig. 30. Lateral view of shrew unicuspids: a) least shrew, b) pygmy shrew, c) southeastern shrew, d) masked shrew, e) water shrew, f) Arctic shrew, and g) smoky shrew

6b. Mental foramen located farther forward; first lower incisor only extends as far back as anterior margin of fourth lower tooth or less (fig. 31c, bottom); greatest width of skull less than 11 mm . . . . .7.

7a (6b). Approaches Great Lakes drainage most closely in Pennsylvania; infraorbital foramen located partly posterior to imaginary line passing between first and second upper molars (third and second teeth, respectively, from the *rear*. Note that rearmost tooth of shrews is small and not readily seen from side) (fig. 31d, top) . . . . Long-tailed Shrew, *Sorex dispar*, p. 31.

7b. Range variable; infraorbital foramen completely anterior to imaginary line passing between first and second molars (third and second teeth, respectively, from the *rear*) (fig. 31d, bottom) . . . . .8.

8a (7b). Greatest length of skull less than 17.5 mm . . . . .9.

8b. Greatest length of skull more than 17.5 mm . . . . .10.

9a (8a). Approaches Great Lakes region in Indiana and Illinois; in lateral view, third unicuspid usually smaller than fourth (fig. 30c); rostrum short and broad (length of unicuspid row less than width measured across fourth unicuspids) . . . . Southeastern Shrew, *Sorex longirostris*, p. 38.

9b. Widespread in Great Lakes region; in lateral view, third unicuspid usually not smaller than fourth (fig. 30d); rostrum longer and more narrow (length of unicuspid row greater than width measured across fourth unicuspids) . . . . Masked Shrew, *Sorex cinereus*, p. 28.

10a (8b). In lateral view, third unicuspid usually smaller than fourth (fig. 30e); greatest length of skull more than 20.5 mm . . . .
. . . . Water Shrew, *Sorex palustris*, p. 40.

10b. In lateral view, third unicuspid usually larger than fourth (fig. 30f–g); greatest length of skull less than 20.5 mm . . . . .11.

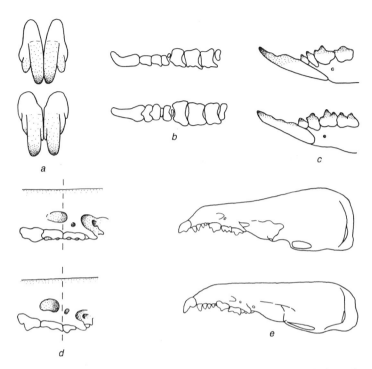

Fig. 31. Characteristics of shrew skulls: a) front view of first incisor of a least shrew (top) compared to a pygmy shrew (bottom), b) occlusal view of upper toothrow of a least shrew (top) compared to a pygmy shrew (bottom), c) side view of mandible of a short-tailed shrew (top) compared to a typical *Sorex* (bottom), d) side view of rostrum and posterior part of toothrow in a long-tailed shrew (top) compared to other *Sorex* (bottom) (anterior is to the left), and e) side view of skull of an Arctic shrew (top) compared to a smoky shrew (bottom)

11a (10b). Greatest width of skull 9.2 mm or more; dorsal outline of skull not flat in lateral view, somewhat inflated in cranial region (fig. 31e, top) . . . .
. . . . Arctic Shrew, *Sorex arcticus*, p. 26.
11b. Greatest width of skull 9.3 mm or less; dorsal outline of skull usually flat in lateral view (fig. 31e, bottom) . . . . Smoky Shrew, *Sorex fumeus*, p. 33.

III. Order Chiroptera: Family Vespertilionidae, Plain-nosed Bats.
1a. One upper incisor on each side . . . . 2.
1b. Two upper incisors on each side . . . . 4.

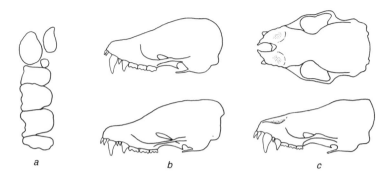

Fig. 32. Characteristics of bat skulls: a) occlusal view of upper teeth of a red bat, b) side view of toothrow in a silver-haired bat or eastern pipistrelle (top) compared to a *Myotis* (bottom), and c) dorsal and lateral view of skull of a silver-haired bat

2a (1a). Five upper cheek teeth (teeth behind canine) on each side, first minute and hidden from lateral view (fig. 32a); base of incisor contacts canine . . . . 3.

2b. Four upper cheek teeth on each side, first well formed, easily visible in lateral view; base of incisor not in contact with canine . . . .
. . . . Evening Bat, *Nycticeius humeralis,* p. 82.

3a (2a). Greatest length of skull 15 mm or less . . . .
. . . . Red Bat, *Lasiurus borealis,* p. 74.

3b. Greatest length of skull 16 mm or more. . . . .
. . . . Hoary Bat, *Lasiurus cinereus,* p. 77.

4a (1b). Four upper cheek teeth on each side; greatest length of skull 17.5 mm or more. Big Brown Bat, *Eptesicus fuscus,* p. 88, fig. 6.

4b. More than four upper cheek teeth on each side; greatest length of skull less than 17.5 mm . . . . 5.

5a (4b). Five upper cheek teeth on each side; in lateral view, first upper cheek tooth much smaller than second (fig. 32b, top). . . . . 6.

5b. Six upper cheek teeth on each side; in lateral view, first and second upper cheek teeth about equal in size and both much smaller than third (fig. 32b, bottom). . . . . 7.

6a (5a). Shallow depression extending forward and medially from each orbit to nostril opening (fig. 32c); six lower cheek teeth on each side; greatest length of skull more than 14 mm. . . . .
. . . . Silver-haired Bat, *Lasionycteris noctivagans,* p. 80.

6b. No such depressions; five lower cheek teeth on each side; greatest length of skull less than 14 mm . . . . Eastern Pipistrelle, *Pipistrellus subflavus,* p. 85.

7a (5b). Greatest length of skull usually less than 14.1 mm; least interorbital width less than 3.4 mm . . . . Small-footed Bat, *Myotis leibii,* p. 63.

7b. Greatest length of skull usually greater than 14.1 mm; least interorbital width greater than 3.4 mm . . . . 8.
8a (7b). Least interorbital width usually 4 mm or more; greatest width of braincase generally 7.3 mm or more. . . . . Little Brown Bat, *Myotis lucifugus*, p. 66.
8b. Least interorbital width usually less than 4 mm; greatest width of braincase generally 7.3 mm or less . . . . 9.
9a (8b). Rostrum broad, such that length of maxillary tooth row (from front of canine to back of last cheek tooth) slightly less than width across last cheek teeth; ratio of maxillary tooth row length to least interorbital width less than 1.5 . . . . Indiana Bat, *Myotis sodalis*, p. 71.
9b. Rostrum more narrow, such that length of maxillary tooth row greater than width across last cheek teeth; ratio of maxillary tooth row length to least interorbital width more than 1.5 . . . . Northern Bat, *Myotis septentrionalis*, p. 69.

IV. Order Lagomorpha: Family Leporidae, Rabbits, and Hares.

1a. Raised shield-shaped area on midline above foramen magnum (supraoccipital region) usually higher than wide (width measured at midpoint) (fig. 33a); boundaries of interparietal bone usually not discernible (fig. 33c); posterior border of supraorbital process generally not touching cranium (fig. 33c) . . . 2.
1b. Raised shield-shaped area on midline above foramen magnum usually wider than high (fig. 33b); boundaries of interparietal bone usually discernible (fig. 33d–e); posterior border of supraorbital process may or may not contact cranium (fig. 33d–e). . . . . 4.
2a (1a). Greatest length of skull less than 88 mm . . . . Snowshoe Hare, *Lepus americanus*, p. 99.
2b. Greatest length of skull more than 88 mm . . . . 3.
3a (2b). Found in New York, Ontario, or Michigan; length of nasal bones usually more than 42 mm; ratio of nasal bone length to greatest length of skull greater than 0.43 . . . . European Hare, *Lepus europaeus*, p. 102.
3b. Found in Wisconsin or Minnesota; length of nasal bones usually less than 42 mm; ratio of nasal bone length to greatest length of skull less than 0.43 . . . . White-tailed Jackrabbit, *Lepus townsendii*, p. 105.
4a (1b). Posterior tip of supraorbital process usually touching cranium, often fused with cranium (fig. 33d); boundary between frontal and nasal bone smooth or slightly roughened (fig. 33d); anterior extension of supraorbital process usually obvious (fig. 33d); widely distributed in Great Lakes region . . . . Eastern Cottontail, *Sylvilagus floridanus*, p. 94, fig. 7.
4b. Posterior tip of supraorbital process closely approaches but often does not contact cranium (fig. 33e); boundary between frontal and nasal bone very jagged (fig. 33e); anterior extension of supraorbital process absent or barely

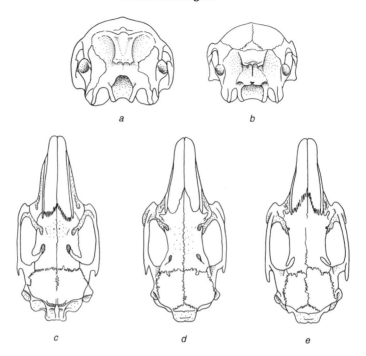

Fig. 33. Characteristics of rabbit and hare skulls: a) occipital region of a hare, b) occipital region of a rabbit, c) dorsal view of the skull of a hare, d) dorsal view of the skull of an eastern cottontail, and e) dorsal view of the skull of an Appalachian cottontail

discernible (fig. 33e); approaches southeastern boundary of Great Lakes region . . . . Appalachian Cottontail, *Sylvilagus obscurus*, p. 96.

V. Order Rodentia: Rodents.

1a. Ear opening at end of distinct, upward curving tube; skull large, greatest length 115 mm or more. . . . .
. . . . **Family Castoridae:** American Beaver, *Castor canadensis*, p. 142, fig. 10.

1b. Ear opening lateral; upward curving auditory tube lacking; skull smaller, greatest length less than 115 mm . . . . 2.

2a (1b). Infraorbital foramen as large or larger than foramen magnum; skull large, total length usually 95 mm or larger . . . . **Family Erethizontidae:** Common Porcupine, *Erethizon dorsatum*, p. 197, fig. 14.

2b. Infraorbital foramen smaller than foramen magnum; skull smaller, total length usually less than 95 mm. . . . . 3.

3a (2b). Postorbital process present on frontal bone, sharply pointed...........
.... **Family Sciuridae: Squirrels** . . . 4.
3b. No postorbital process present on frontal bone...................... 13.
4a (3a). Top of skull almost flat in outline; incisor usually white, occasionally pale yellow; postorbital process at 90° angle to long axis of skull...........
.... Woodchuck, *Marmota monax*, p. 115.
4b. Top of skull gently curved in outline; incisors yellow or orange; postorbital process points back at angle less than 90° to long axis of skull........... 5.
5a (4b). Zygomatic arch twisted, midpoint approximately horizontal, definitely at less than 45° angle; width of skull just posterior to postorbital process greater than least interorbital width .... 6.
5b. Midpoint of zygomatic arch closer to vertical, definitely at greater than 45° angle; width of skull just posterior to postorbital process not noticeably greater than least interorbital width... 7.
6a (5a). Greatest length of skull 41 mm or less ....
..... Thirteen-lined Ground Squirrel, *Spermophilus tridecemlineatus*, p. 120.
6b. Greatest length of skull greater than 41 mm ....
.... Franklin's Ground Squirrel, *Spermophilus franklinii*, p. 118.
7a (5b). In ventral view, infraorbital foramen visible, positioned forward and lateral of first well-formed cheek tooth .... 8.
7b. In ventral view, infraorbital foramen not visible, positioned anterior to first well-formed cheek tooth .... 9.
8a (7a). Five upper cheek teeth on each side, the first small and peglike; greatest length of skull less than 34 mm . . . Least Chipmunk, *Tamias minimus*, p. 110.
8b. Four upper cheek teeth on each side, all well formed; greatest length of skull more than 34 mm .... Eastern Chipmunk, *Tamias striatus*, p. 112.
9a (7b). Greatest length of skull more than 50 mm; auditory bulla with two internal partitions (septa) (fig. 34a, left) that may or may not be externally visible... 10.
9b. Greatest length of skull less than 50 mm; auditory bulla with two or three septa externally visible (fig. 34a)................................... 11.
10a (9a). Five upper cheek teeth on each side; first upper cheek tooth small, peglike .... Eastern Gray Squirrel, *Sciurus carolinensis*, p. 123.
10b. Four upper cheek teeth on each side; all well formed..................
.... Eastern Fox Squirrel, *Sciurus niger*, p. 126, fig. 8.
11a (9b). Auditory bulla with three internal divisions visible externally (fig. 34a, right); greatest length of skull more than 40 mm ....
.... Red Squirrel, *Tamiasciurus hudsonicus*, p. 129.
11b. Auditory bulla with two internal divisions visible externally (fig. 34a, left); greatest length of skull less than 40 mm .... 12.
12a (11b). Greatest length of skull 36 mm or less; greatest width of skull usually

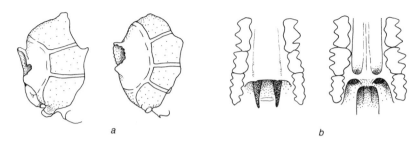

a   *b*

Fig. 34. Characteristics of squirrel and vole skulls: a) ventral view of auditory bullae of squirrels showing two (left) or three (right) partitions, and b) ventral view of the palate of *Clethrionomys* (left) compared to one typical of *Microtus* or *Phenacomys* (right)

21.5 mm or less; length of upper row of cheek teeth usually less than 7 mm . . . . Southern Flying Squirrel, *Glaucomys volans*, p. 134.

12b. Greatest length of skull usually 36 mm or more; greatest width of skull usually greater than 21.5 mm; length of upper row of cheek teeth usually greater than 7 mm . . . . Northern Flying Squirrel, *Glaucomys sabrinus*, p. 131.

13a (3b). Each upper incisor with two longitudinal grooves, one deep and situated on anterior surface, one shallow and located along medial edge; four lower cheek teeth on each side; zygomatic arches widely flared . . . .

Family Geomyidae: Plains Pocket Gopher, *Geomys bursarius*, p. 138, fig. 9.

13b. Upper incisors with zero or one groove; three lower cheek teeth on each side; zygomatic arch variable . . . . 14.

14a (13b). Infraorbital foramen large, always faces forward, oval-shaped such that long axis points down and out; pin-sized opening located directly below infraorbital foramen; upper incisors with single longitudinal groove on anterior surface . . . . **Family Dipodidae: Jumping Mice** . . . 15.

14b. Infraorbital foramen always narrower at bottom, often V-shaped, may face forward or laterally; pin-sized opening below infraorbital foramen missing; upper incisors plain or grooved . . . .

. . . . **Family Muridae: Mice, Rats, Voles, Bog Lemmings, Muskrat** . . . 16.

15a (14a). Four upper cheek teeth, the first small and peglike. . . . .

. . . . Meadow Jumping Mouse, *Zapus hudsonius*, p. 192, fig. 13.

15b. Three upper cheek teeth, all well formed . . . .

. . . . Woodland Jumping Mouse, *Napaeozapus insignis*, p. 189.

16a (14b). Chewing surfaces of cheek teeth with series of whitish enamel loops surrounding patches of dentine, outer margins with zigzag appearance (fig. 35a–h) . . . . 17.

16b. Cheek teeth with low rounded cusps, outer margins without zigzag appearance (fig. 35i–j). . . . . 26.

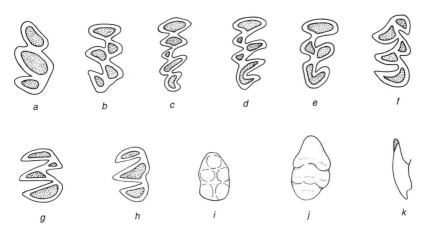

Fig. 35. Rodent teeth: *a*) first upper cheek tooth of an Allegheny woodrat, *b*) first upper cheek tooth of a muskrat, *c*) posterior upper cheek tooth of a rock vole, *d*) posterior upper cheek tooth of a meadow vole, *e*) posterior upper cheek tooth similar to that of a prairie vole, woodland vole, or heather vole, *f*) second lower cheek tooth of a heather vole, *g*) lower cheek tooth of a southern bog lemming, *h*) lower cheek tooth of a northern bog lemming, *i*) cheek tooth of New World mice, *j*) cheek tooth of Old World mice and rats, and *k*) lateral view of upper incisor of a house mouse. All cheek teeth are from the right side and shown in occlusal view

17a (16a). Greatest length of skull more than 35 mm . . . . 18.

17b. Greatest length of skull less than 35 mm . . . . 19.

18a (17a). Greatest length of skull less than 57 mm; first upper cheek tooth with only three patches of dentine (loosely termed *triangles* regardless of actual shape) (fig. 35a); found south of Great Lakes drainage . . . .
. . . . Allegheny Woodrat, *Neotoma magister,* p. 157.

18b. Greatest length of skull 57 mm or more; first upper cheek tooth with more than three triangles (fig. 35b); found throughout Great Lakes drainage . . . .
. . . . Muskrat, *Ondatra zibethicus,* p. 175, fig. 11.

19a (17b). Shallow longitudinal groove on front lateral surface of upper incisor . . . . 20.

19b. No longitudinal groove on front lateral surface of upper incisor . . . . 21.

20a (19a). Posterior edge of palate lacks spine at midpoint; posterior *lower* cheek tooth with three large patches of dentine (loosely termed *triangles* regardless of actual shape) and one smaller triangle along outer edge (fig. 35g); found throughout Great Lakes region . . . .
. . . . Southern Bog Lemming, *Synaptomys cooperi,* p. 180.

20b. Posterior edge of palate with sharp spine at midpoint; posterior *lower* cheek tooth with only three large triangles of dentine, lacking smaller triangle along outer edge (fig. 35h); approaches Great Lakes region in extreme northwest . . . . Northern Bog Lemming, *Synaptomys borealis*, p. 178.

21a (19b). Posterior border of palate shelflike, not supported at midline (fig. 34b, left) . . . . Southern Red-backed Vole, *Clethrionomys gapperi*, p. 159.

21b. Posterior border of palate supported at midline (fig. 34b, right) . . . . 22.

22a (21b). Four indentations (reentrant angles) on either side of posterior upper cheek tooth (fig. 35c) . . . . Rock Vole, *Microtus chrotorrhinus*, p. 165.

22b. Fewer than four reentrant angles on either side of posterior upper cheek tooth . . . . 23.

23a (22b). Three reentrant angles on either side of posterior upper cheek tooth (fig. 35d) . . . . Meadow Vole, *Microtus pennsylvanicus*, p. 170.

23b. Two reentrant angles on either side of posterior upper cheek tooth (fig. 35e) . . . . 24.

24a (23b). Reentrant angles on tongue side of all *lower* cheek teeth much deeper than outside angles (fig. 35f); found only in northern Great Lakes basin . . . . Heather Vole, *Phenacomys intermedius*, p. 162.

24b. Reentrant angles on both sides of first two *lower* cheek teeth approximately equal; found in central or southern Great Lakes basin. . . . . 25.

25a (24b). Least interorbital width usually 4.3 mm or more; foramen above opening to ear usually less than 3 mm long; ratio of greatest length of skull to greatest width less than 1.7 . . . . Woodland Vole, *Microtus pinetorum*, p. 173.

25b. Least interorbital width usually less than 4.3 mm; foramen above opening to ear usually more than 3 mm long; ratio of greatest length of skull to greatest width more than 1.7 . . . . Prairie Vole, *Microtus ochrogaster*, p. 167.

26a (16b). Cheek teeth with two longitudinal rows of cusps at midtooth (fig. 35i) . . . . 27.

26b. Cheek teeth with three longitudinal rows of cusps at midtooth (fig. 35j) . . . . 29.

27a (26a). Incisors with longitudinal groove on front lateral surface . . . . Western Harvest Mouse, *Reithrodontomys megalotis*, p. 148.

27b. Incisors without longitudinal groove on front lateral surface . . . . 28.

28a (27b). Greatest length of skull less than 24.8 mm . . . . Deer Mouse (grassland form), *Peromyscus maniculatus bairdii*, p. 153.

28b. Greatest length of skull usually 24.8 mm or more . . . . other subspecies of Deer Mouse, *Peromyscus maniculatus*, p. 153, or the White-footed Mouse, *Peromyscus leucopus*, p. 151, fig. 12. (There are no simple characteristics that reliably differentiate the skulls of these species.)

29a (26b). Greatest length of skull less than 25 mm; in lateral view, tip of upper incisor distinctly notched (fig. 35k) . . . . House Mouse, *Mus musculus*, p. 183.

29b. Greatest length of skull more than 25 mm; in lateral view, tip of upper incisor not notched . . . . Norway Rat, *Rattus norvegicus*, p. 186.

VI. Order Carnivora: Carnivores.
1a. Last upper cheek tooth about 1.5 times as long as wide, long axis parallels long axis of skull, larger than adjacent tooth; skull massive, greatest length exceeds 250 mm . . . .
. . . . **Family Ursidae:** Black Bear, *Ursus americanus*, p. 214, fig. 16.
1b. Last upper cheek tooth variable in shape, if obviously elongate then long axis points laterally at 45–90° angle; last upper cheek tooth usually not larger than adjacent tooth; greatest length of skull usually less than 250 mm . . . . 2.
2a (1b). Three or four upper cheek teeth on each side; posterior upper cheek tooth very small, oriented with long axis perpendicular to long axis of adjacent tooth; three lower cheek teeth present on each side, all bladelike (lacking flat crushing surfaces); rostrum short . . . . **Family Felidae: Cats** . . . 3.
2b. Four or more upper cheek teeth on each side; posterior upper cheek tooth not tiny relative to adjacent tooth, not oriented as above; five or more lower cheek teeth on each side, not all bladelike; rostrum variable . . . . 6.
3a (2a). Four upper cheek teeth on each side . . . . 4.
3b. Three upper cheek teeth on each side . . . . 5.
4a (3a). Greatest length of skull more than 150 mm . . . .
. . . . Mountain Lion, *Felis concolor*, p. 274.
4b. Greatest length of skull less than 100 mm . . . . Domestic Cat, *Felis catus*.
5a (3b). One large opening between auditory bullae and foramen magnum (fig. 36a); maximum width of presphenoid bone less than 6 mm (fig. 36a) . . . .
. . . . Bobcat, *Lynx rufus*, p. 254, fig. 19.
5b. Two foramina (hypoglossal canal and jugular foramen) opening separately between auditory bullae and foramen magnum (fig. 36b); maximum width of presphenoid bone 6 mm or more (fig. 36b) . . . .
. . . . Canadian Lynx, *Lynx canadensis*, p. 252.
(Some domestic cats also have only three upper cheek teeth on each side. However, bobcat and lynx skulls are more than 100 mm in greatest length, whereas a domestic cat skull is always shorter than this.)
6a (2b). Six upper cheek teeth on each side . . . . 7.
6b. Four or five upper cheek teeth on each side . . . .
. . . . **Family Mustelidae: Weasels, Skunks, Badger, and Otter** . . . 13.
7a (6a). Posterior border of palate well behind last upper tooth (more than 20% of palate is behind last tooth); six lower cheek teeth . . . .
. . . . . **Family Procyonidae:** Common Raccoon, *Procyon lotor*, p. 218, fig. 17.
7b. Posterior border of palate even with last tooth or slightly behind (less than 20% of palate is behind last tooth); seven lower cheek teeth . . . .
. . . . **Family Canidae: Dogs** . . . 8.
8a (7b). Slight longitudinal depression in frontal bone at base of each postorbital process (fig. 37) . . . . 9.

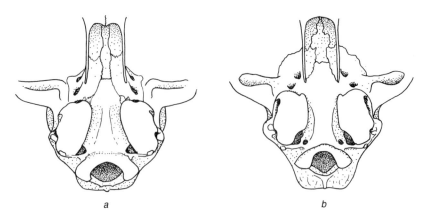

Fig. 36. Ventral view of bobcat (a) and Canadian lynx (b) skulls

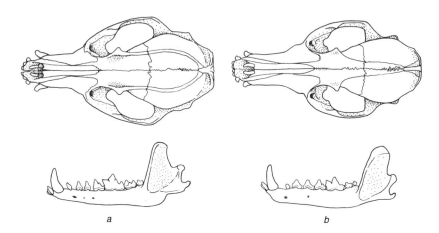

Fig. 37. Common gray fox (a) and red fox (b) skulls and mandibles

8b. Base of postorbital processes smooth or convex, no paired depressions . . . . .
. . . . 10.

9a (8a). Temporal ridges forming broad U when viewed from above (fig. 37a),
distance between center of ridges (measured at junction of frontal and parietal
bones) greater than 18 mm; lower part of mandible notched at posterior end
(fig. 37a) . . . . Common Gray Fox, *Urocyon cinereoargenteus*, p. 211.

9b. Temporal ridges most often forming a narrow V (fig. 37b), separated by 18
mm or less at junction of frontal and parietal bones; lower part of mandible
not notched at posterior end (fig. 37b) . . . . . . Red Fox, *Vulpes vulpes*, p. 208.

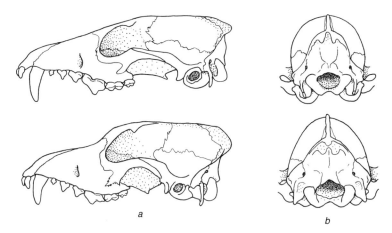

Fig. 38. Separating coyote (top) and domestic dog (bottom) skulls: a) lateral view and b) posterior view

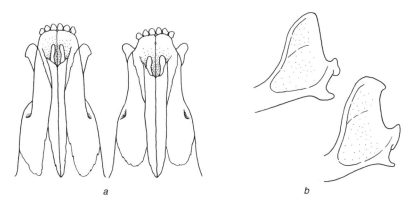

Fig. 39. Separating gray wolf and domestic dog skulls: a) dorsal view of the rostrum in a gray wolf (left) and a domestic dog (right), and b) medial view of the posterior mandible in a gray wolf (left) and a domestic dog (right)

10a (8b). Greatest length of skull less than 225 mm . . . . 11.
10b. Greatest length of skull 225 mm or more. . . . . 12.
11a (10a). Forehead gently rising from rostrum (fig. 38a, top); maximum width of braincase located near suture between parietal and squamosal bones (fig. 38b, top); narrow extension of frontal bone parallels nasal bone for more than 30% of nasal bone length (length of nasal bone measured as longest straight-line distance between anterior and posterior margins; similar to that shown

in fig. 39a, left); area between postorbital processes not obviously swollen; auditory bulla smooth, rounded . . . . Coyote, *Canis latrans*, p. 202.

11b. Forehead rising abruptly from rostrum (fig. 38a, bottom); maximum width of braincase usually lower, near base of zygomatic arch (fig. 38b, bottom); anterior extension of frontal bone usually shorter and wider, generally less than 30% of nasal bone length (see fig. 39a, right); area above eye swollen; auditory bulla usually rough (rugose) and somewhat flattened, especially lateral surface . . . . Domestic Dog, *Canis familiaris*. (There is immense variation in the sizes and proportions of the skulls of domestic dogs, and coyote-dog hybrids exist. Positive identification might require comparison to known material in a museum.)

12a (10b). Narrow extension of frontal bone (fig. 39a, left) parallels nasal bone for more than 30% of its length (length of nasal bone measured as longest straight-line distance between anterior and posterior margins); posterior border of coronoid process of mandible usually straight (fig. 39b, left); palate usually ends at rear boundary of last upper tooth or farther forward; forehead gently rises from rostrum or with moderate bulge . . . .
. . . . Gray Wolf, *Canis lupus*, p. 206.

12b. Anterior extension of frontal bone usually wider and shorter (fig. 39a, right), generally less than 30% of length of nasal bone; posterior border of coronoid process of mandible often curves backward at top (fig. 39b, right); palate occasionally ends at rear border of last upper tooth but often extends back for another 5–15 mm; forehead rises abruptly from rostrum . . . .
. . . . Domestic Dog, *Canis familiaris*. (There is immense variation in the sizes and proportions of the skulls of domestic dogs, and wolf-dog hybrids exist. Positive identification might require comparison to known material in a museum.)

13a (6b). Five upper cheek teeth on each side . . . . 14.

13b. Four upper cheek teeth on each side . . . . 17.

14a (13a). Five lower cheek teeth; infraorbital foramen large, about 8 mm long and 5 mm wide or larger . . . . Northern River Otter, *Lutra canadensis*, p. 247.

14b. Six lower cheek teeth; infraorbital foramen smaller, usually less than 6 mm long and 4 mm wide . . . . .15.

15a (14b). Rostrum broader than long; greatest length of skull 130 mm or more. . . . . Wolverine, *Gulo gulo*, p. 273.

15b. Rostrum longer than broad; greatest length of skull less than 130 mm . . . . .
. . . . .16.

16a (15b). Exposed rootlet usually present on outer side of fourth upper cheek tooth (upper carnassial) (fig. 40e); greatest length of skull more than 95 mm . . . . Fisher, *Martes pennanti*, p. 226.

16b. No exposed rootlet on outer side of fourth upper cheek tooth; greatest length of skull 95 mm or less. . . . . American Marten, *Martes americana*, p. 223.

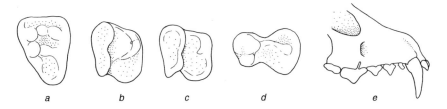

Fig. 40. Carnivore teeth: occlusal views of the posterior cheek tooth in a) American badger, b) striped skunk, c) spotted skunk, d) weasel or mink (*Mustela*); and e) side view of the rostrum of a fisher showing exposed rootlet

17a (13b). Last upper cheek tooth triangular in outline (fig. 40a); skull roughly triangular in outline when viewed from above; greatest length of skull greater than 100 mm . . . . American Badger, *Taxidea taxus*, p. 239.

17b. Last upper cheek tooth not triangular; skull not triangular; greatest length of skull usually less than 80 mm . . . . 18.

18a (17b). Posterior border of palate approximately even with posterior border of last upper tooth; anterior border of last upper tooth straight, not noticeably indented (fig. 40b–c) . . . . 19.

18b. Posterior border of palate well behind posterior border of last upper tooth; midpoint of anterior and posterior border of last upper tooth indented, resulting in hourglass shape to chewing surface (fig. 40d) . . . . 20.

19a (18a). When viewed from side, top of skull a continuous curve; greatest length of skull usually greater than 60 mm . . . .
. . . . Striped Skunk, *Mephitis mephitis*, p. 244, fig. 18.

19b. When viewed from side, top of skull mostly straight; greatest length of skull less than 60 mm . . . . Spotted Skunk, *Spilogale putorius*, p. 242.

20a (18b). Greatest length of skull more than 55 mm . . . .
. . . . Mink, *Mustela vison*, p. 236.

20b. Greatest length of skull less than 55 mm . . . . 21.

21a (20b). Greatest length of skull 35 mm or less . . . .
. . . . Least Weasel, *Mustela nivalis*, p. 234.

21b. Greatest length of skull usually more than 35 mm . . . . 22.

22a (21b). Postglenoid length of skull more than 46% of greatest length in males, more than 48% of greatest length in females; greatest width of skull less than 20 mm in females, less than 24 mm in males; postorbital process blunt, weakly developed . . . . Ermine, *Mustela erminea*, p. 228.

22b. Postglenoid length of skull less than 46% of greatest length in males, less than 48% of greatest length in females; greatest width of skull usually 20 mm or more in females, usually 24 mm or more in males; postorbital process more

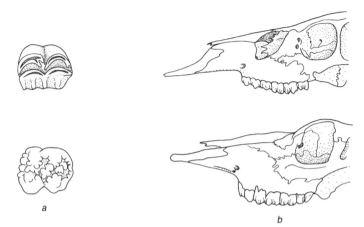

Fig. 41. Characteristics of deer, cow, and pig skulls: *a*) occlusal views of a cheek tooth of a cervid (top) compared to a pig (bottom), and *b*) lateral view of the rostrum of a cervid (top) compared to a bovid (bottom)

sharply pointed, better developed . . . .
. . . . Long-tailed Weasel, *Mustela frenata*, p. 231.

VII. Order Artiodactyla: Deer, Bison, Cattle, Sheep, Goat, and Pig.

1a. Upper incisors lacking; six upper cheek teeth on each side; cheek teeth with elongate or crescent-shaped ridges of enamel (selenodont), not rounded cusps (fig. 41a, top); orbit surrounded by complete, circular, bony ring . . . . 2.

1b. Upper incisors present; seven upper cheek teeth on each side; cheek teeth with many rounded cusps (fig. 41a, bottom); orbit not surrounded by complete, circular, bony ring . . . . **Family Suidae: Domestic Pig,** *Sus scrofa.*

2a (1a). Lacrimal bone contacts nasal bone (fig. 41b, bottom); usually only one opening (lacrimal foramen) along anterior margin of orbit (may be just inside, outside, or directly on rim of orbit); rostrum not fenestrated to great degree anterior to orbit; horns may be present, never antlers . . . .
. . . . **Family Bovidae: Bison, Cattle, Sheep, and Goat** . . . 3.

2b. Lacrimal bone does not contact nasal bone (fig. 41b, top); two openings along anterior margin of orbit; rostrum with large openings anterior to orbit; antlers may be present, never horns . . . . **Family Cervidae: Deer** . . . 6.

3a (2a). Top of skull obviously slopes down and back from highest point near orbit, rear boundary much lower than highest point along midline; greatest length of skull less than 250 mm . . . . 4.

3b. Top of skull reasonably flat such that posterior margin is about as high as

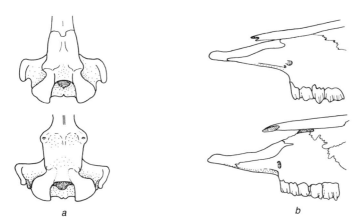

Fig. 42. Separating goat from sheep and cattle from bison skulls: a) ventral view of occipital area of a domestic goat (top) compared to domestic sheep (bottom), and b) lateral view of the rostrum of domestic cattle (top) compared to American bison (bottom)

midline between orbits; greatest length of skull usually much greater than 250 mm . . . . 5.

4a (3a). Region anterior to foramen magnum on ventral surface (basioccipital area) wider in back than front (fig. 42a, top); premaxillary bone usually in broad contact with nasal bone; paired depressions anterior to orbit either missing or small. . . . . Domestic Goat, *Capra hircus*.

4b. Region anterior to foramen magnum on ventral surface roughly rectangular or slightly wider in front than back (fig. 42a, bottom); premaxillary bone barely touches or not in contact with nasal bone; obvious depression anterior to orbit on either side, about 12–18 mm wide, 18–24 mm long. . . . . Domestic Sheep, *Ovis aries*.

5a (3b). When viewed from above, zygomatic arch totally hidden by overlapping frontal bone; dorsal process of premaxillary bone never contacts nasal bone, usually does not approach within 30 mm of nasal bone (fig. 42b, bottom) . . . . American Bison, *Bison bison*, p. 275.

5b. When viewed from above, zygomatic arch mostly visible, not hidden by overlapping frontal bone; dorsal process of premaxillary bone extends up to within 10 mm of nasal bone, often making contact (fig. 42b, top). . . . . Domestic Cattle, *Bos taurus*.

6a (2b). Nasal bones short, not extending forward of most anterior upper cheek tooth; if antler present, main beam palmate, ultimately curving backward (fig. 43a). . . . . Moose, *Alces alces*, p. 267.

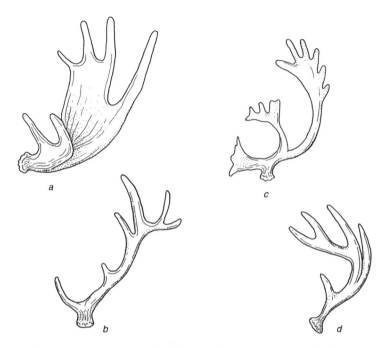

Fig. 43. Antlers: a) moose, b) elk, c) caribou, and d) white-tailed deer

6b. Nasal bones longer, extending forward of most anterior upper cheek tooth; antlers varied . . . . 7.

7a (6b). When viewed from below, nasal cavity at rear margin of palate separated into right and left halves by thin vertical sheet of bone (part of the vomer); if antler present, main beam ultimately curves forward. . . . . 8.

7b. When viewed from below, nasal cavity at rear margin of palate not separated into right and left halves; if antler present, main beam cylindrical, ultimately curving backward (fig. 43b) . . . . Wapiti or Elk, *Cervus elaphus*, p. 261.

8a (7a). Greatest length of skull more than 350 mm; small upper canines may be visible; antlers may be present in either sex, somewhat palmate (fig. 43c) . . . . Caribou, *Rangifer tarandus*, p. 270.

8b. Greatest length of skull less than 350 mm; upper canines absent; antlers present only in males, cylindrical, never palmate (fig. 43d) . . . . White-tailed Deer, *Odocoileus virginianus*, p. 264.

VIII. Perissodactyla: Family Equidae, Horses and Asses . . . . Horse, *Equus caballus*.

Key to Adult Mammals in the Flesh

Key to Orders

1a. Hallux (big toe) thumblike, pointing inward, opposable, and not clawed; snout conical; nose pinkish; outer ear very thin, paperlike; tail round and naked; body size equal to small house cat; abdominal pouch (marsupium) present in females . . . . . . **New World Opossums: Order Didelphimorphia, I.**

1b. Not possessing above combination of characters . . . . 2.

2a (1b). Forelimbs modified as wings . . . . **Bats: Order Chiroptera, III.**

2b. Forelimbs adapted for running, walking, digging, or swimming . . . . 3.

3a (2b). Toes surrounded by hooves; large bodied, total length greater than 1,250 mm . . . . **Deer: Order Artiodactyla, VII.**

3b. Toes with nails or claws, not hooves; total length variable . . . . 4.

4a (3b). Tail shorter than ear, well furred, tuftlike; hindfoot longer than tail, 2.5–3.0 times longer than forefoot . . . .
. . . . **Rabbits and Hares: Order Lagomorpha, IV.**

4b. Not possessing above combination of characters . . . . 5.

5a (4b). Body with cat- or doglike appearance; four toes on hindfoot . . . .
. . . . **Carnivores, in part (Wild Cats and Dogs): Order Carnivora, VI.**

5b. Body not cat- or doglike; five toes on hindfoot . . . . 6.

6a (5b). Four well-formed toes with claws on forefoot (a vestigial fifth toe with a nail, not a claw, may be visible on inside of foot). . . . . **Rodents, in part (excluding Muskrat, Beaver, and Pocket Gopher): Order Rodentia, V.**

6b. Five well-formed toes with claws on forefoot . . . . 7.

7a (6b). Tail naked *and* total length greater than 225 mm *and* either hindfoot webbed or external cheek pouches present . . . .
Rodents, in part (Muskrat, Beaver, and Pocket Gopher): Order Rodentia, V.

7b. Not possessing above combination of characters . . . . 8.

8a (7b). External ear (pinna) minute or absent; total length less than 210 mm . . .
. . . . **Insectivores (Shrews and Moles): Order Insectivora, II.**

8b. Pinna well formed; total length always greater than 170 mm, generally greater than 225 mm . . . . **Carnivores, in part (Weasels, Skunks, Raccoon, and Bear): Order Carnivora, VI.**

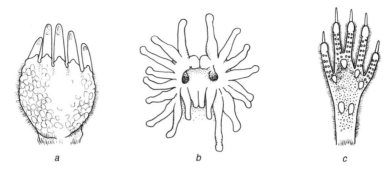

Fig. 44. Characteristics of insectivores: *a*) the forefoot of an eastern mole, *b*) front view of the snout of a star-nosed mole, *c*) bottom surface of the rear foot of a water shrew

Key to Species by Order

I. Order Didelphimorphia: Family Didelphidae, New World Opossums. . . . . Virginia Opossum, *Didelphis virginiana*, p. 19.

II. Order Insectivora: Shrews and Moles.

1a. Forefoot large, two or more times as wide as hindfoot, tipped with massive claws (fig. 44a); longest claw on forefoot represents more than 25% of forefoot length; palms face outward; ear pinna missing; total length greater than 145 mm . . . . **Family Talpidae: Moles** . . . 2.

1b. Forefoot not obviously enlarged, similar in proportions to hindfoot; longest claw on forefoot less than 25% of forefoot length; palms face downward; ear pinna small, often hidden by fur; total length usually less than 145 mm . . . . **Family Soricidae: Shrews** . . . 4.

2a (1a). Snout tipped with 22 fleshy tentacles (fig. 44b); nostrils face forward . . . . Star-nosed Mole, *Condylura cristata*, p. 56.

2b. Snout lacking fleshy tentacles; nostrils not facing forward . . . . 3.

3a (2b). Nostrils open on side of snout; coarse hairs cover tail, many 5 mm or more in length; feet not webbed; hindfoot length less than 21 mm . . . . Hairy-tailed Mole, *Parascalops breweri*, p. 51.

3b. Nostrils open on top of snout; tail sparsely furred or naked; all feet webbed; hindfoot length greater than 21 mm . . . . Eastern Mole, *Scalopus aquaticus*, p. 54.

4a (1b). Tail length less than 30% of total length . . . . 5.

4b. Tail length greater than 30% of total length . . . . 6.

5a (4a). Total length less than 95 mm; found only in southern Great Lakes region . . . . Least Shrew, *Cryptotis parva*, p. 43.

5b. Total length greater than 95 mm; found throughout Great Lakes region . . . . Northern Short-tailed Shrew, *Blarina brevicauda,* p. 46.

6a (4b). Feet and toes with fringe of whitish hairs (fig. 44c); hindfoot length greater than 16 mm; total length greater than 130 mm . . . .
. . . . Water Shrew, *Sorex palustris,* p. 40.

6b. Feet and toes without fringe of whitish hairs; hindfoot length less than 16 mm; total length less than 130 mm . . . . 7.

7a (6b). Total length greater than 100 mm . . . . 8.

7b. Total length less than or equal to 100 mm . . . . 10.

8a (7a). Distinct line of demarcation between dark back and lighter sides; body often tricolored (darkest on back, sides intermediate, lightest on ventral surface) . . . . Arctic Shrew, *Sorex arcticus,* p. 26.

8b. Color of back same as sides or grades into that of sides; body not tricolored . . . . 9.

9a (8b). Tail length usually greater than 52 mm, more than 75% of head and body length; approaches eastern boundary of Great Lakes region . . . .
. . . . Long-tailed Shrew, *Sorex dispar,* p. 31.

9b. Tail length usually 52 mm or less, less than 75% of head and body length; widespread over eastern portion of Great Lakes region. . . . .
. . . . Smoky Shrew, *Sorex fumeus,* p. 33.

10a (7b). Only three well-formed unicuspids visible in lateral view (see fig. 30b); hindfoot length usually 10 mm or less . . . . . . Pygmy Shrew, *Sorex hoyi,* p. 36.

10b. Four unicuspids visible in lateral view; hindfoot length usually 10 mm or more. . . . . 11.

11a (10b). Third unicuspid equal to or larger than fourth (see fig. 30d); tail length usually greater than 31 mm; widely distributed in Great Lakes region . . . . Masked Shrew, *Sorex cinereus,* p. 28.

11b. Third unicuspid usually smaller than fourth (see fig. 30c); tail length usually 31 mm or less; approaches Great Lakes region most closely in Indiana and Illinois. . . . . Southeastern Shrew, *Sorex longirostris,* p. 36.

III. Order Chiroptera: Family Vespertilionidae, Plain-nosed Bats.

1a. Fur with at least some patches of white-tipped hairs giving animal a frosted appearance . . . . 2.

1b. Fur lacking white-tipped hairs . . . . 4.

2a (1a). Overall fur color brick red to yellowish red . . . .
. . . . Red Bat, *Lasiurus borealis,* p. 74.

2b. Overall fur color black, brown, or tan . . . . 3.

3a (2b). Yellowish brown or tan fur; dorsal surface of tail membrane completely furred; total length greater than 120 mm; forearm length greater than 44 mm . . . . Hoary Bat, *Lasiurus cinereus,* p. 77.

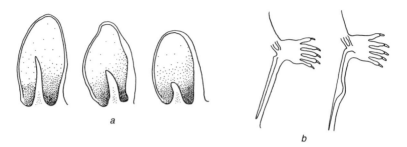

Fig. 45. Bat ears and calcars: *a*) ear of a northern bat (left), other *Myotis* (center), and an evening bat (right), and *b*) a calcar without a keel (left) compared to a calcar with a keel (right)

3b. Fur dark brown or black; dorsal surface of tail membrane only furred over proximal 25–50%; total length less than 120 mm; forearm length less than 44 mm . . . . Silver-haired Bat, *Lasionycteris noctivagans*, p. 80.

4a (1b). Forearm length greater than 40 mm; total length greater than 105 mm . . . . Big Brown Bat, *Eptesicus fuscus*, p. 88.

4b. Forearm length 40 mm or less; total length less than 105 mm . . . . 5.

5a (4b). Body hairs obviously tricolored—black at base, yellowish brown along shaft, and dark brown or black at tip; proximal third of dorsal surface of tail membrane lightly furred; dorsal surface of feet and toes with light-brown hairs sharply contrasting with black color of underlying skin . . . . Eastern Pipistrelle, *Pipistrellus subflavus*, p. 85.

5b. Body hairs not obviously tricolored; color of hairs on feet and toes not sharply contrasting with color of underlying skin; tail membrane not furred to any extent . . . . 6.

6a (5b). Tragus of ear slender, sharply pointed (fig. 45a, left), long (about 8–9 mm); when laid forward, ear extends 3 mm or more past tip of nose; ear height 16 mm or more. . . . . Northern Bat, *Myotis septentrionalis*, p. 69.

6b. Tragus variable in shape but not sharply pointed, less than 8 mm in height; when laid forward, ear extends less than 2 mm past tip of nose; ear height 16 mm or less . . . . 7.

7a (6b). Light-brown fur with distinct black facial mask; hindfoot length 8 mm or less; forearm length less than 34 mm; total length less than 85 mm; calcar keeled (fig. 45b, right) . . . . Small-footed Bat, *Myotis leibii*, p. 63.

7b. No obvious facial mask; hindfoot length 8 mm or more; forearm length greater than 34 mm; total length greater than 85 mm; calcar variable . . . . . 8.

8a (7b). Tragus broadly rounded, often curved (fig. 45a, right), short, usually 5 mm or less in height; only one upper incisor on each side. . . . . Evening Bat, *Nycticeius humeralis*, p. 82.

8b. Tragus bluntly pointed (fig. 45a, middle), greater than 5 mm in height; two upper incisors on each side . . . . 9.

9a (8b). Calcar strongly keeled (fig. 45b, right); hairs between toes not extending beyond tip of toes; fur dull, usually dark gray or black . . . . Indiana Bat, *Myotis sodalis*, p. 71.

9b. Calcar not obviously keeled (fig. 45b, left); hairs between toes extending beyond tip of toes; fur glossy, usually brownish . . . . Little Brown Bat, *Myotis lucifugus*, p. 66.

IV. Order Lagomorpha: Family Leporidae, Rabbits and Hares.

1a. Nape with rusty or cinnamon brown patch; hindfoot length less than 115 mm . . . . 2.

1b. Nape without rusty or cinnamon brown patch; hindfoot length greater than 115 mm . . . . 3.

2a (1a). Black spot between ears; white forehead blaze never present; approaches southeastern boundary of Great Lakes region . . . . Appalachian Cottontail, *Sylvilagus obscurus*, p. 96.

2b. Lacking black spot between ears (although a dark spot may be present lower down on forehead); white forehead blaze may be present; found throughout southern Great Lakes region . . . . Eastern Cottontail, *Sylvilagus floridanus*, p. 94.

3a (1b). Tail length less than 50 mm; body of summer specimens brownish, dorsal hairs slate gray at base; tail of summer specimens black or slate gray above, slightly lighter in color but still somewhat dark below; winter specimens appear white all over except for ear tips . . . . Snowshoe Hare, *Lepus americanus*, p. 99.

3b. Tail length greater than 50 mm; body of summer specimens brownish, dorsal hairs whitish or light gray at base; tail of summer specimens variable above, white or light gray below; winter specimens may or may not turn white . . . 4.

4a (3b). Tail of summer specimens white with at most a faint black or gray stripe down dorsal midline; dorsal hairs of summer specimens dirty white or light gray at base with prominent cinnamon-brown band before tan tip; winter specimens white except ear tips . . . . White-tailed Jackrabbit, *Lepus townsendii*, p. 105.

4b. Tail of summer specimens completely black above and snow white below; dorsal hairs of summer specimens snow white at base with distinct black band before brownish tip; winter and summer coats identical . . . . European Hare, *Lepus europaeus*, p. 102.

V. Order Rodentia: Rodents.

1a. Some hairs modified as sharp-pointed quills . . . . **Family Erethizontidae:** Common Porcupine, *Erethizon dorsatum*, p. 197.

1b. No hairs modified as quills. . . . . .2.

2a (1b). External cheek pouches present, opening in anterior direction, lateral to mouth . . . .

. . . . . . **Family Geomyidae:** Plains Pocket Gopher, *Geomys bursarius,* p. 138.

2b. External cheek pouches not present . . . . .3.

3a (2b). Tail naked, paddle shaped, dorsoventrally flattened such that width much greater than 25% of length; total length greater than 650 mm. . . . .

. . . . **Family Castoridae:** American Beaver, *Castor canadensis,* p. 142.

3b. Tail naked or furred, not paddle shaped, not flattened as above; total length less than 650 mm . . . . .4.

4a (3b). Hindfeet large, about 2.5 times as long as forefeet; tail round, scaly, longer than head and body; total length less than 260 mm . . . .

. . . .**Family Dipodidae: Jumping Mice** . . . 5.

4b. Hindfeet generally smaller, usually less than 2.5 times as long as forefeet; tail and total length variable . . . . .6.

5a (4a). Terminal 5 mm or more of tail white; lateral fur brightly colored, often yellowish orange . . . .

. . . . Woodland Jumping Mouse, *Napaeozapus insignis,* p. 189.

5b. Tail tip dark, similar in color to proximal portion; lateral fur more drab, generally olive brown or yellowish brown . . . .

. . . . Meadow Jumping Mouse, *Zapus hudsonius,* p. 192.

6a (4b). Tail well furred, often bushy; length of longest tail hairs much greater than width of tail vertebrae, always greater than 10 mm . . . .

. . . . **Family Sciuridae: Squirrels** . . . 7.

6b. Tail naked or furred, never bushy; length of longest tail hairs generally less than width of tail vertebrae, always less than 10 mm . . . .

. . . . **Family Muridae: Mice, Rats, Voles, Bog Lemmings, Muskrat** . . . 16.

7a (6a). Furred flap of skin stretching between wrist and ankle alongside body . . .

. . . . .8.

7b. No flap of skin between wrist and ankle. . . . . .9.

8a (7a). Belly hairs slate colored at base. . . . .

. . . . Northern Flying Squirrel, *Glaucomys sabrinus,* p. 131.

8b. Belly hairs completely white . . . .

. . . . Southern Flying Squirrel, *Glaucomys volans,* p. 134.

9a (7b). Alternating light and dark stripes on back. . . . . .10.

9b. No alternating light and dark stripes on back . . . . .12.

10a (9a). Six solid light-colored stripes alternating with seven dark-brown stripes; each brown stripe contains row of light-colored rectangles . . . .

. . . . . Thirteen-lined Ground Squirrel, *Spermophilus tridecemlineatus,* p. 120.

10b. Five dark brown stripes separated by four relatively light-colored stripes (light brown medially and white laterally); no light-colored rectangles. . . . 11.

11a (10b). Middle stripes obviously extend across rump to tail; no reddish brown rump patch . . . . Least Chipmunk, *Tamias minimus*, p. 110.

11b. No stripes extend across rump; rump reddish brown. . . . .
. . . . Eastern Chipmunk, *Tamias striatus*, p. 112.

12a (9b). Claws on middle digits of forefoot long, maximum length obviously greater than 50% of length of fleshy portion of digits. . . . . .13.

12b. Claws on middle digits of forefoot short, maximum length about 50% the length of fleshy portion of digits or less . . . . .14.

13a (12a). Feet black or dark brown; many hairs on back tipped with white or silver, resulting in a frosted appearance; body large; total length greater than 425 mm . . . . Woodchuck, *Marmota monax*, p. 115.

13b. Feet light brown or gray; hairs on back not tipped with white or silver; body medium; total length less than 425 mm . . . .
. . . . Franklin's Ground Squirrel, *Spermophilus franklinii*, p. 118.

14a (12b). Winter specimens with reddish tint along back, black ear tufts present; summer specimens less red, thin black stripe separates white underparts from dark sides, no ear tufts present; total length less than 375 mm . . . .
. . . . Red Squirrel, *Tamiasciurus hudsonicus*, p. 129.

14b. No black ear tufts or black side stripe; total length greater than 375 mm . . .
. . . . .15.

15a (14b). Body either brown with grayish cast or totally black; if brown with grayish cast, belly white and hair tips on tail gray or silver. . . . .
. . . . Eastern Gray Squirrel, *Sciurus carolinensis*, p. 123.

15b. Body with overall brownish or orangish appearance; belly and hair tips on tail orange (fulvous). . . . . Eastern Fox Squirrel, *Sciurus niger*, p. 126.

16a (6b). Hindfeet webbed; tail flattened from side to side (laterally compressed), naked; total length greater than 450 mm . . . .
. . . . Muskrat, *Ondatra zibethicus*, p. 175.

16b. Hindfeet not webbed; tail not laterally compressed, naked or furred; total length less than 450 mm . . . . .17.

17a (16b). Tail similar in length to hindfoot, not more than 5 mm greater in length than hindfoot. . . . . .18.

17b. Tail obviously longer than hindfoot, more than 5 mm greater in length than hindfoot . . . . .20.

18a (17a). Fur with reddish tint (auburn or chestnut) along back and sides; dorsal hairs molelike, most less than 10 mm in length; incisors without longitudinal grooves . . . . Woodland Vole, *Microtus pinetorum*, p. 173.

18b. Fur a grizzled brown to black along back and sides, somewhat shaggy; many dorsal hairs greater than 10 mm in length; incisors with longitudinal grooves. . . . . .19.

19a (18b). Found throughout Great Lakes region; six mammae . . . .
. . . . Southern Bog Lemming, *Synaptomys cooperi*, p. 180.

19b. Approaches Great Lakes region in extreme northwest; eight mammae . . . . .
. . . . Northern Bog Lemming, *Synaptomys borealis*, p. 178.
(Positive identification of bog lemmings found in the northwestern Great Lakes
basin may require an examination of the skull.)

20a (17b). Reddish brown head and back contrasting with grayish brown
sides . . . . Southern Red-backed Vole, *Clethrionomys gapperi*, p. 159.

20b. Lacking reddish brown band on head and neck . . . . 21.

21a (20b). Tail length less than 50% of head and body length; ear pinna mostly
hidden by fur. . . . . 22.

21b. Tail length greater than 50% of head and body length; ear pinna prominent,
not mostly hidden by fur . . . . 25.

22a (21a). Snout yellowish or orangish in color. . . . . 23.

22b. Snout not yellowish or orangish in color . . . . 24.

23a (22a). Tail length greater than 40 mm . . . .
. . . . Rock Vole, *Microtus chrotorrhinus*, p. 165.

23b. Tail length less than 40 mm . . . .
. . . . Heather Vole, *Phenacomys intermedius*, p. 162.

24a (22b). Belly grayish or silvery; six plantar tubercles on hindfoot; eight mammae. Meadow Vole, *Microtus pennsylvanicus*, p. 170.

24b. Belly yellowish brown or tan; five plantar tubercles on hindfoot; six mammae . . . . Prairie Vole, *Microtus ochrogaster*, p. 167.

25a (21b). Tips of most belly hairs snow white or creamy white . . . . 26.

25b. Tips of most belly hairs gray, tan, or brown. . . . . 30.

26a (25a). Tail length greater than 125 mm; found south of Great Lakes drainage . . . . Allegheny Woodrat, *Neotoma magister*, p. 157.

26b. Tail length less than 125 mm; range variable . . . . 27.

27a (26b). Sides yellowish or orangish brown, somewhat grizzled; small yellowish or orangish brown patch occasionally in front of ears; incisors with
longitudinal groove; found only in southwestern portion of Great Lakes region . . . . Western Harvest Mouse, *Reithrodontomys megalotis*, p. 148.

27b. Sides light brown to reddish brown; no colored patch in front of ears;
incisors without longitudinal groove; widely distributed. . . . . 28.

28a (27b). Hindfoot usually 19 mm or less; ear height usually 15 mm or less; tail
length usually less than 65 mm, much less than head and body length; tail
most often distinctly bicolored with terminal tuft of white hairs; usually grassland habitats; found south of line running from central Wisconsin, eastward
across Michigan's Thumb and Ontario to southern shore of Lake Ontario. . . .
. . . . . . Deer Mouse (grassland form), *Peromyscus maniculatus bairdii*, p. 153.

28b. Hindfoot usually 19 mm or more; ear height usually 16 mm or more; tail
length usually greater than 65 mm; tail, habitat, and range variable . . . . . . 29.

29a (28b). Ear height 18 mm or less; tail most often indistinctly bicolored, gradually changing from dark on top to light below; white tail tuft usually missing;

throat hairs completely white . . . .
. . . . White-footed Mouse, *Peromyscus leucopus*, p. 151.

29b. Ear height 18 mm or more; tail most often distinctly bicolored, abruptly changing from dark above to light below; white tail tuft usually present; throat hairs often gray at base . . . .
. . . . other subspecies of Deer Mouse, *Peromyscus maniculatus*, p. 153.

30a (25b). Total length less than 200 mm . . . .
. . . . House Mouse, *Mus musculus*, p. 183.

30b. Total length greater than 200 mm . . . .
. . . . Norway Rat, *Rattus norvegicus*, p. 186.

VI. Order Carnivora: Weasels, Bears, Raccoons, Cats, and Dogs.

1a. Body with cat- or doglike appearance; four toes on hindfoot . . . . 2.

1b. Body not cat- or doglike; five toes on hindfoot . . . . .7.

2a (1a). Body catlike, at least twice the size of a house cat; claws curved, sharp, retractile; lacking scent gland at dorsal base of tail; tail short, not bushy, less than 200 mm in length . . . . **Family Felidae: Wild Cats** . . . 3.

2b. Body doglike; claws straight, somewhat blunt, not retractile; scent gland on back at dorsal base of tail; tail long and bushy, greater than 200 mm in length . . . . **Family Canidae: Wild Dogs** . . . 4.

3a (2a). Ears with tuft of black hairs about 40 mm in length; black marking at tip of tail is continuous around tail; hindfoot length greater than 200 mm . . . . . .
. . . . Canadian Lynx, *Lynx canadensis*, p. 252.

3b. Ear tufts lacking or very small; black marking at tip of tail on dorsal surface only; hindfoot length less than 200 mm . . . . Bobcat, *Lynx rufus*, p. 254.

4a (2b). Tail length greater than 33% of total length; total length less than 1,075 mm . . . . .5.

4b. Tail length less than 33% of total length; total length greater than 1,075 mm . . . . .6.

5a (4a). Body usually reddish (occasionally brownish or black); feet black; tail tipped with white . . . . Red Fox, *Vulpes vulpes*, p. 208.

5b. Body mostly a mix of gray and black; feet not black; tail tipped with black . . . . Common Gray Fox, *Urocyon cinereoargenteus*, p. 211.

6a (4b). Total length greater than 1,350 mm; hindfoot length greater than 225 mm; width of nosepad greater than 25 mm . . . .
. . . . Gray Wolf, *Canis lupus*, p. 206.

6b. Total length less than 1,350 mm; hindfoot length less than 225 mm; width of nosepad less than 25 mm. . . . . Coyote, *Canis latrans*, p. 202.

7a (1b). Body massive, not elongate, uniform black in color both above and below (except possible white throat patch); total length usually much greater than 800 mm; tail inconspicuous, less than 125 mm in length . . . .
. . . . **Family Ursidae:** Black Bear, *Ursus americanus*, p. 214.

7b. Body shape variable, color never uniform black above and below; total length usually less than 800 mm (if greater than 800 mm then feet webbed); tail obvious, length variable .... 8.

8a (7b). Tail with multiple dark rings; black facial mask ....
.... **Family Procyonidae:** Common Raccoon, *Procyon lotor,* p. 218.

8b. Tail without multiple black rings; no black facial mask..................
.... **Family Mustelidae: Weasels, Skunks, Badger, and Otter** ... 9.

9a (8b). White stripes and/or spots on head and/or back; body not particularly thin or elongate... 10.

9b. No white stripes or spots on head or back; body thin, elongate .... 12.

10a (9a). Front claws massive; single white stripe down midline from snout to shoulders; no other stripes or spots on back; white face with black markings........................... American Badger, *Taxidea taxus,* p. 239.

10b. Front claws not greatly enlarged compared to claws on rear feet; multiple white stripes or spots on back; black face with variable white markings... 11.

11a (10b). At most, two broad, white stripes, extending back from shoulders; top of head and neck white .... Striped Skunk, *Mephitis mephitis,* p. 244.

11b. Four to six white stripes, beginning on shoulders and breaking into spots or blotches on back; top of head and neck black, except for white spots or blotches..................... Spotted Skunk, *Spilogale putorius,* p. 242.

12a (9b). Ventral fur mostly white or yellowish white; total length less than 450 mm... 13.

12b. Ventral fur dark, may have light-colored blotches; total length usually greater than 450 mm .... 15.

13a (12a). Terminal portion of tail black............................... 14.

13b. Terminal portion of tail not obviously black (a few black hairs may be present)......................... Least Weasel, *Mustela nivalis,* p. 234.

14a (13a). Tail length more than 44% of head and body length ....
.... Long-tailed Weasel, *Mustela frenata,* p. 231.

14b. Tail length less than 44% of head and body length ....
.... Ermine, *Mustela erminea,* p. 228.

15a (12b). Ears 25 mm or less in height, without light-colored border .... 16.

15b. Ears more than 25 mm in height, with light-colored border .... 17.

16a (15a). Feet fully webbed; tail unusually thick and muscular at base, strongly tapering toward tip; total length greater than 700 mm ....
.... Northern River Otter, *Lutra canadensis,* p. 247.

16b. Feet partly webbed; tail not unusually thick and muscular at base, tapering less obvious; total length less than 700 mm .... Mink, *Mustela vison,* p. 236.

17a (15b). Body yellowish brown to dark brown; large orange, buff, or cream-colored patch covering throat and extending onto chest; total length 660 mm or less .... American Marten, *Martes americana,* p. 223.

17b. Body dark brown to black; no large buff or cream-colored patch covering throat and extending onto chest, although small patches may occur on throat and/or chest; total length usually much greater than 660 mm . . . .
. . . . Fisher, *Martes pennanti,* p. 226.

VII. Order Artiodactyla: Family Cervidae, Deer.
1a. Snout broad, overhangs upper lip; loose flap of skin hangs from neck; shoulders humped; antlers present in males only; if antlers present, main beam palmate, ultimately curving backward (see fig. 43a) . . . .
. . . . Moose, *Alces alces,* p. 267.
1b. Not possessing above combination of characters . . . . 2.
2a (1b). Neck and mane yellowish white, lighter than head and back; patch of white on leg above hooves; antlers may be present in either sex; palmate brow tine of antler extends over eye (see fig. 43c). . . . .
. . . . Caribou, *Rangifer tarandus,* p. 270.
2b. Not colored as above; antlers present only in males, not palmate . . . . 3.
3a (2b). Head and neck dark brown, much darker than back and sides; chin and throat brown; if antlers present, main beam cylindrical, ultimately curving backward (see fig. 43b) . . . . Wapiti or Elk, *Cervus elaphus,* p. 261.
3b. Head and neck mostly brown, similar in color to back and sides; chin and throat white; if antlers present, main beam cylindrical, ultimately curving forward (see fig. 43d) . . . . White-tailed Deer, *Odocoileus virginianus,* p. 264.

Summary of Measurements and Life History Information

Species	Total Length (mm)	Tail Length (mm)	Hindfoot Length (mm)	Ear Height (mm)	Weight (g, unless kg indicated)	Skull Length (mm)
Marsupials						
Virginia opossum	650–850	250–380	50–80	40–60	2.0–5.5kg	105–126
Insectivores						
Arctic shrew	105–125	37–46	12–15	—	7–13	18.7–20.1
Masked shrew	75–100	31–45	10–13	—	3.5–5.5	16.0–17.5
Long-tailed shrew	110–135	50–64	14–15	—	4–6	17.3–18.5
Smoky shrew	110–126	42–52	12–15	—	6–11	18.0–19.2
Pigmy shrew	80–91	27–33	8.0–10.5	—	2–4	15.0–16.5
Southeastern shrew	77–92	26–32	9–11	—	3–6	14.1–15.9
Water shrew	138–164	63–74	19–20	—	10–18	21.1–22.5
Least shrew	64–86	12–18	9–12	—	4.0–6.5	16.0–17.2
Short-tailed shrew	108–140	18–32	13–18	—	15–30	20.8–24.8
Hairy-tailed mole	150–170	23–36	18–20	—	40–64	31–34
Eastern mole	150–200	20–38	22–27	—	65–140	35–40
Star-nosed mole	175–205	65–85	25–30	—	35–75	33–35
Bats						
Small-footed bat	73–82	30–35	7–8	12–15	5–7	13.2–14.2
Little brown bat	80–95	31–45	9–11	13–16	6–12	14.1–15.4
Northern bat	77–92	34–45	8–10	16–19	5–8	14.5–15.5
Indiana bat	73–100	27–44	8–10	12–15	6–11	14.1–15.0
Red bat	93–117	40–50	6–11	9–13	7–13	13.1–14.6
Hoary bat	130–150	53–64	10–14	17–19	25–35	17.0–18.5
Silver-haired bat	90–115	35–50	6–12	15–17	8–12	16.1–17.0
Evening bat	86–103	33–40	8–10	11–14	6–12	14.4–15.1
Eastern pipistrelle	70–80	28–45	7–10	13–15	4–7	12.5–13.3
Big brown bat	110–130	38–50	10–14	16–20	15–24	18.5–20.7
Rabbits and Hares						
Eastern cottontail	375–475	35–70	80–110	50–65	900–1,800	69–81
Appalachian cottontail	386–430	22–65	87–96	54–63	800–1,000	69–76
Snowshoe hare	380–505	25–45	120–150	60–70	1.4–2.3kg	75–82
European hare	640–700	70–100	145–160	79–100	3.0–5.6kg	92–105
White-tailed jackrabbit	560–650	70–100	135–160	95–110	2.5–4.5kg	88–103
Rodents						
Least chipmunk	185–222	80–100	28–35	13–18	42–53	31–33
Eastern chipmunk	225–266	65–110	32–40	12–20	66–115	38–42
Woodchuck	530–650	100–160	65–95	25–40	2.3–5.0kg	73–95
Franklin's ground squirrel	350–420	125–160	50–57	16–18	370–500	52–57
Thirteen-lined ground squirrel	225–300	75–109	32–41	7–11	100–220	37–41
Eastern gray squirrel	425–500	180–250	60–74	25–35	350–700	59–64
Eastern fox squirrel	500–590	220–270	62–80	22–32	700–1,100	63–70
Red squirrel	280–345	100–145	40–55	18–26	135–250	44–47
Northern flying squirrel	245–310	110–150	35–40	18–26	70–130	35–39
Southern flying squirrel	220–257	85–115	26–33	13–18	50–75	33–36
Plains pocket gopher	235–310	63–90	30–37	6–9	200–400	47–55
American beaver	900–1,200	300–400	170–195	30–35	12–27kg	119–140
Western harvest mouse	120–152	51–70	14–17	10–12	8–15	19.0–21.4
White-footed mouse	145–195	65–95	19–23	15–18	16–30	24.1–27.4

Skull Width (mm)	Number of Teeth	Number of Toes (forefoot-hindfoot)	Mammae	Gestation (days)	Litters per Year	Litter Size	Birth Weight (g, unless kg indicated)	Record Longevity (years)
56–70	50	5–5	9–17	13	1–2	7–9	0.15	6
9.2–10.1	32	5–5	6	20–23?	1–3	4–10	—	1.5
7.4–8.4	32	5–5	6	18	1–3	5–7	0.25	1.5?
7.9–8.1	32	5–5	6	—	—	2–5	—	—
8.5–9.3	32	5–5	6	20	1–3	4–6	—	1.5?
6.1–7.4	32	5–5	6	18?	1	3–8	—	1.5?
7.0–7.8	32	5–5	6	18?	2?	1–6	—	—
10.0–10.9	32	5–5	6	21?	2–3	5–7	—	1.5?
7.5–8.3	30	5–5	6	21–23	2–3	3–6	0.3	1.75
11.3–13.0	32	5–5	6	21–22	1–3	4–7	—	2.75
14–15	44	5–5	8	28–42?	1	3–6	—	4
18–21	36	5–5	6	40–45	1	3–5	5.4	2
13.0–14.2	44	5–5	8	45?	1	3–7	1.5	—
8.0–9.0	38	5–5	2	60?	1	1	—	12
8.6–9.6	38	5–5	2	60	1	1	2.3	33
8.9–9.4	38	5–5	2	50–60	1	1	—	19
8.4–9.3	38	5–5	2	60?	1	1	—	14
9.0–10.4	32	5–5	4	60–90?	1	2–3	1.5?	—
11.8–12.9	32	5–5	4	60–90?	1	2	4.5	—
9.6–10.4	36	5–5	2	50–60	1	2	2.0	12
9.8–10.4	30	5–5	2	50–60?	1	2	2.0	5
7.5–8.2	34	5–5	2	44–60?	1	2	1.6	15
11.8–13.5	32	5–5	2	60	1	2	3.3	19
34–40	28	5–4	8	28	2–4	3–8	25–30	4
35–37	28	5–4	8	28	2–4	3–8	—	—
37–42	28	5–4	8	37	2–4	3–6	65–80	—
45–50	28	5–4	8	40–42	2	3–5	130	—
45–52	28	5–4	8	42	2–4	3–7	90	8
16–19	22	4–5	8	30	1	4–7	2.2	—
21–24	20	4–5	8	31	2	2–5	3	8
49–60	22	4–5	8	32	1	2–5	26	6
30–33	22	4–5	10	26–28	1	4–11	—	—
20.7–25.4	22	4–5	10	28	1–2	6–11	6–7	—
32–37	22	4–5	8	44	2	2–4	15–18	15
35–40	20	4–5	8	44	2	2–5	15	15
24–28	20	4–5	8	31–35	1–2	3–6	6.7	10
21.2–23.4	22	4–5	8	37–42	1	2–4	5–6	4
19.6–21.6	22	4–5	8	40	2	2–5	3–5	13
26–36	20	5–5	6	51?	1	2–6	5	7
80–98	20	5–5	4	107	1	2–5	450	21
10.0–11.0	16	4–5	6	23–24	2–4	2–6	1–1.5	1?
12.2–14.1	16	4–5	6	22–23	4–5	2–7	2	2

Species	Total Length (mm)	Tail Length (mm)	Hindfoot Length (mm)	Ear Height (mm)	Weight (g, unless kg indicated)	Skull Length (mm)
Deer mouse (*P. m. bairdii*)	120–165	42–69	15–19	12–15	10–24	22.4–24.7
Deer mouse (other subspecies)	160–205	70–110	18–23	18–21	12–24	24.7–27.8
Allegheny woodrat	360–440	156–200	39–45	25–28	260–450	45–56
Southern red-backed vole	120–160	31–50	17–21	12–16	15–35	22.5–25.1
Heather vole	135–155	25–40	17–19	12–16	30–45	23.0–26.5
Rock vole	140–185	42–60	18–22	13–17	30–45	25.0–27.7
Prairie vole	125–155	30–40	18–22	11–14	30–60	24.7–28.9
Meadow vole	130–185	35–60	18–25	11–16	35–60	25.8–28.8
Woodland vole	110–140	18–24	16–19	10–13	20–40	21.7–25.4
Muskrat	470–630	200–270	70–90	20–25	800–1,500	61–69
Northern bog lemming	110–140	18–26	17–21	12–14	20–50	22–27
Southern bog lemming	110–140	18–24	16–20	10–13	20–45	24.3–27.1
House mouse	150–190	70–95	17–21	11–18	15–23	20.4–22.5
Norway rat	320–450	125–190	30–45	16–20	200–490	42–45
Woodland jumping mouse	210–250	120–155	28–33	16–18	19–32	22.7–25.2
Meadow jumping mouse	180–225	110–140	28–34	12–16	12–28	20.8–23.4
Common porcupine	600–900	160–220	75–110	25–40	5–14kg	92–111
Carnivores						
Coyote	1,100–1,300	290–390	180–220	95–125	11–21kg	175–208
Gray wolf	1,350–1,700	370–475	235–295	100–130	35–65kg	232–263
Red fox	950–1,050	325–400	140–175	75–90	3.5–7.0kg	134–156
Common gray fox	875–1,050	300–390	125–145	70–80	3.5–6.8kg	119–130
Black bear	1,250–1,800	80–125	200–280	100–135	110–230kg	256–288
Common raccoon	700–925	220–260	110–125	50–60	6–20kg	112–125
American marten (♂)	605–660	195–220	90–95	35–45	1.0–1.5kg	80–89
American marten (♀)	550–570	170–200	75–84	34–38	725–1,000	72–75
Fisher (♂)	910–1,075	335–400	110–125	41–55	3.0–5.5kg	115–122
Fisher (♀)	800–880	280–350	90–115	40–49	2–3kg	98–106
Ermine (♂)	277–330	73–95	37–45	16–22	90–155	39–46
Ermine (♀)	235–252	50–69	28–33	15–21	50–73	34–38
Long-tailed weasel (♂)	350–440	111–150	38–50	16–23	170–270	44–51
Long-tailed weasel (♀)	290–338	85–122	30–40	15–22	85–125	35–44
Least weasel (♂)	189–210	30–44	21–23	12–15	41–60	31–33
Least weasel (♀)	170–182	22–31	19–21	12–15	30–45	30–31
Mink (♂)	550–680	185–220	62–75	17–25	900–1,300	63–71
Mink (♀)	480–550	140–185	50–65	15–23	600–900	58–65
American badger	700–800	115–150	95–140	45–60	6–11kg	110–131
Spotted skunk	470–540	180–230	45–50	25–30	375–600	43–55
Striped skunk	550–675	175–280	60–80	25–35	1.5–5.0kg	66–82
Northern river otter	900–1,300	320–500	110–135	10–25	5–14kg	100–115
Canadian lynx	825–1,050	75–125	205–250	70–80	5–17kg	122–146
Bobcat	750–1,100	130–180	160–195	60–85	5–16kg	115–141
Deer						
Elk	2,100–2,700	120–160	600–655	140–200	240–450kg	415–520
White-tailed deer	1,550–2,100	250–350	475–525	140–225	45–140kg	270–330
Moose	2,000–2,800	80–120	730–835	255–265	330–500kg	540–590
Caribou	1,650–2,300	100–150	500–660	110–150	90–180kg	395–400

Skull Width (mm)	Number of Teeth	Number of Toes (forefoot-hindfoot)	Mammae	Gestation (days)	Litters per Year	Litter Size	Birth Weight (g, unless kg indicated)	Record Longevity (years)
11.3–12.8	16	4–5	6	22–23	4	2–7	1.5	8
12.3–13.9	16	4–5	6	22–23	4	2–7	1.5	8
23.5–29.0	16	4–5	4	35	2–3	2–4	15	3.5
12.0–13.7	16	4–5	8	17–19	2–3	3–6	1.9	1.75
12.8–15.5	16	4–5	8	19–24	1–3	2–8	2.0–2.7	—
14.2–16.0	16	4–5	8	19–21?	2–3?	1–7	—	—
13.8–15.9	16	4–5	6	19–21	3–5	2–6	2.9	3
14.3–15.7	16	4–5	8	21	4–8	4–8	2–3	1
13.1–15.8	16	4–5	4	20–24	1–4	2–4	2–3	1+
37–43	16	5–5	6	25–30	1–3	4–8	21	4?
13.5–18.5	16	4–5	8	—	—	2–8	—	—
14.3–16.7	16	4–5	6	23–26	2–3	2–5	4	2.5
10.3–11.5	16	4–5	10	19–21	5–10	4–8	1	6
20.5–22.1	16	4–5	12	21	6–8	6–11	6	3
11.7–12.9	16	4–5	8	23–29	2	2–7	1	4?
10.0–11.3	18	4–5	8	18	2	2–8	0.8	5
62–79	20	4–5	4	210	1	1	400–500	10
92–110	42	5–4	8	58–65	1	3–8	260	18
122–143	42	5–4	10	63	1	5–7	500	16
69–79	42	5–4	8	51–54	1	4–6	100	6
66–74	42	5–4	6	53?	1	3–5	86	10
149–174	42	5–5	6	210	1	1–3	275	26
70–80	40	5–5	6	63	1	2–6	85	22
46–53	38	5–5	—	—	—	—	—	—
39–45	38	5–5	4	217–277	1	1–4	28	17
64–74	38	5–5	—	—	—	—	—	—
52–61	38	5–5	4	330–360	1	1–4	40	10
20.3–23.5	34	5–5	—	—	—	—	—	—
16.0–19.1	34	5–5	8–10	270–300	1	4–8	2	7
23.5–27.8	34	5–5	—	—	—	—	—	—
19.7–23.0	34	5–5	8	240–300	1	4–8	3	3
15.8–17.5	34	5–5	—	—	—	—	—	—
14.7–15.9	34	5–5	8	35	2–3	3–7	1.5	—
35–42	34	5–5	—	—	—	—	—	—
32–41	34	5–5	8	40–75	1	4–9	6	10
77–89	34	5–5	8	210?	1	2–5	94	26
32–36	34	5–5	10	50–65	1	4–9	10	10
41–52	34	5–5	10–15	59–77	1	4–6	34	10
66–79	36	5–5	4	270–380	1	2–3	130	20
92–108	28	5–4	4	63–70	1	2–5	200	24
79–105	28	5–4	4	60	1	1–4	325	32
185–220	34	4–4	4	250	1	1	16kg	20
105–145	32	4–4	4	200	1	1–2	2–3kg	20
210–230	32	4–4	4	240	1	1–2	11–16kg	27
180–190	32–34	4–4	4	228	1	1	5–9kg	20

Dental Formulae

One of the easiest ways to identify many skulls is to look at the number and kinds of teeth. For example, the meadow jumping mouse is the only mammal in the Great Lakes basin that has a total of 18 teeth, and the evening bat is the only bat species with 30 teeth. Often one does not have to count the total number of teeth in order to distinguish between similar species. An elk, for instance, has upper canines that a white-tailed deer lacks, and a mountain lion has a total of six upper premolars compared to only four in a lynx.

How does one distinguish between incisors, canines, premolars, and molars? Shape is sometimes helpful, but the best clue is the location of the tooth in the jaw. The incisors are the only teeth rooted in the premaxillary bone, and the lower incisors are those teeth that occlude (come together) with the upper incisors. The upper canine is always rooted in the maxillary bone and appears at the junction of the premaxillary and maxillary bones; when the jaws are closed, the lower canine is positioned directly in front of the upper canine. Behind the canines are the premolars and molars. Although premolars always are anterior to molars, the real distinction between the two types of cheek teeth is developmental. During its lifetime, a typical mammal has two sets (deciduous and adult) of incisors, canines, and premolars, but only one complement of molars, so molars simply are cheek teeth that are never replaced. Fortunately, it often is not necessary to distinguish between premolars and molars in order to identify a skull; it generally is sufficient to know the total number of cheek teeth.

Although one could memorize the number and kinds of teeth possessed by each species, most mammalogists rely on a table of dental formulae, similar to the one below. Look at the first entry for *Didelphis,*

341

the Virginia opossum. The upper case letters in the formula refer to the four types of teeth—I(ncisors), C(anines), P(remolars), and M(olars), and the numerals on the top and bottom of each horizontal line refer to the number of each tooth type in the upper and lower jaws, respectively. The formula indicates, for example, that the opossum has five upper incisors and four lower incisors on each side. After the first equal sign is the total number of teeth in the upper jaw (26) and lower jaw (24), and the final number indicates the total number of teeth in the skull and mandible combined (50). Following each formula is an alphabetical list of the genera, and occasionally species within genera, displaying that particular dental pattern.

$$I\frac{5\text{-}5}{4\text{-}4},C\frac{1\text{-}1}{1\text{-}1},P\frac{3\text{-}3}{3\text{-}3},M\frac{4\text{-}4}{4\text{-}4}=\frac{26}{24}=50 \quad Didelphis$$

$$I\frac{3\text{-}3}{3\text{-}3},C\frac{1\text{-}1}{1\text{-}1},P\frac{4\text{-}4}{4\text{-}4},M\frac{3\text{-}3}{3\text{-}3}=\frac{22}{22}=44 \quad Condylura, Parascalops, Sus$$

$$I\frac{3\text{-}3}{3\text{-}3},C\frac{1\text{-}1}{1\text{-}1},P\frac{4\text{-}4}{4\text{-}4},M\frac{2\text{-}2}{3\text{-}3}=\frac{20}{22}=42 \quad Canis, Urocyon, Ursus \text{ (typical)}, Vulpes$$

$$I\frac{3\text{-}3}{3\text{-}3},C\frac{1\text{-}1}{1\text{-}1},P\frac{4\text{-}4}{3\text{-}3},M\frac{3\text{-}3}{3\text{-}3}=\frac{22}{20}=42 \quad Equus \text{ (highly variable)}$$

$$I\frac{3\text{-}3}{3\text{-}3},C\frac{1\text{-}1}{1\text{-}1},P\frac{4\text{-}4}{4\text{-}4},M\frac{2\text{-}2}{2\text{-}2}=\frac{20}{20}=40 \quad Procyon$$

$$I\frac{2\text{-}2}{3\text{-}3},C\frac{1\text{-}1}{1\text{-}1},P\frac{3\text{-}3}{3\text{-}3},M\frac{3\text{-}3}{3\text{-}3}=\frac{18}{20}=38 \quad Myotis$$

$$I\frac{3\text{-}3}{3\text{-}3},C\frac{1\text{-}1}{1\text{-}1},P\frac{4\text{-}4}{4\text{-}4},M\frac{1\text{-}1}{2\text{-}2}=\frac{18}{20}=38 \quad Gulo, Martes$$

$$I\frac{3\text{-}3}{2\text{-}2},C\frac{1\text{-}1}{0\text{-}0},P\frac{3\text{-}3}{3\text{-}3},M\frac{3\text{-}3}{3\text{-}3}=\frac{20}{16}=36 \quad Scalopus$$

$$I\frac{2\text{-}2}{3\text{-}3},C\frac{1\text{-}1}{1\text{-}1},P\frac{2\text{-}2}{3\text{-}3},M\frac{3\text{-}3}{3\text{-}3}=\frac{16}{20}=36 \quad Lasionycteris$$

$$I\frac{3\text{-}3}{3\text{-}3},C\frac{1\text{-}1}{1\text{-}1},P\frac{4\text{-}4}{3\text{-}3},M\frac{1\text{-}1}{2\text{-}2}=\frac{18}{18}=36 \quad Lutra$$

$$I\frac{2\text{-}2}{3\text{-}3},C\frac{1\text{-}1}{1\text{-}1},P\frac{2\text{-}2}{2\text{-}2},M\frac{3\text{-}3}{3\text{-}3}=\frac{16}{18}=34 \quad Pipistrellus$$

$$I\frac{3\text{-}3}{3\text{-}3},C\frac{1\text{-}1}{1\text{-}1},P\frac{3\text{-}3}{3\text{-}3},M\frac{1\text{-}1}{2\text{-}2}=\frac{16}{18}=34 \quad Mephitis, Mustela, Spilogale, Taxidea$$

$I \dfrac{0\text{-}0}{3\text{-}3}$, $C \dfrac{1\text{-}1}{1\text{-}1}$, $P \dfrac{3\text{-}3}{3\text{-}3}$, $M \dfrac{3\text{-}3}{3\text{-}3} = \dfrac{14}{20} = 34$ Cervus, Rangifer (typical)

$I \dfrac{3\text{-}3}{1\text{-}1}$, $C \dfrac{1\text{-}1}{1\text{-}1}$, $P \dfrac{3\text{-}3}{1\text{-}1}$, $M \dfrac{3\text{-}3}{3\text{-}3} = \dfrac{20}{12} = 32$ Blarina, Sorex

$I \dfrac{2\text{-}2}{3\text{-}3}$, $C \dfrac{1\text{-}1}{1\text{-}1}$, $P \dfrac{1\text{-}1}{2\text{-}2}$, $M \dfrac{3\text{-}3}{3\text{-}3} = \dfrac{14}{18} = 32$ Eptesicus

$I \dfrac{1\text{-}1}{3\text{-}3}$, $C \dfrac{1\text{-}1}{1\text{-}1}$, $P \dfrac{2\text{-}2}{2\text{-}2}$, $M \dfrac{3\text{-}3}{3\text{-}3} = \dfrac{14}{18} = 32$ Lasiurus

$I \dfrac{2\text{-}2}{2\text{-}2}$, $C \dfrac{1\text{-}1}{1\text{-}1}$, $P \dfrac{2\text{-}2}{2\text{-}2}$, $M \dfrac{3\text{-}3}{3\text{-}3} = \dfrac{16}{16} = 32$ Homo

$I \dfrac{0\text{-}0}{3\text{-}3}$, $C \dfrac{0\text{-}0}{1\text{-}1}$, $P \dfrac{3\text{-}3}{3\text{-}3}$, $M \dfrac{3\text{-}3}{3\text{-}3} = \dfrac{12}{20} = 32$ Alces, Bison, Bos, Capra, Odocoileus, Ovis

$I \dfrac{3\text{-}3}{1\text{-}1}$, $C \dfrac{1\text{-}1}{1\text{-}1}$, $P \dfrac{2\text{-}2}{1\text{-}1}$, $M \dfrac{3\text{-}3}{3\text{-}3} = \dfrac{18}{12} = 30$ Cryptotis

$I \dfrac{1\text{-}1}{3\text{-}3}$, $C \dfrac{1\text{-}1}{1\text{-}1}$, $P \dfrac{1\text{-}1}{2\text{-}2}$, $M \dfrac{3\text{-}3}{3\text{-}3} = \dfrac{12}{18} = 30$ Nycticeius

$I \dfrac{3\text{-}3}{3\text{-}3}$, $C \dfrac{1\text{-}1}{1\text{-}1}$, $P \dfrac{3\text{-}3}{2\text{-}2}$, $M \dfrac{1\text{-}1}{1\text{-}1} = \dfrac{16}{14} = 30$ Felis

$I \dfrac{3\text{-}3}{3\text{-}3}$, $C \dfrac{1\text{-}1}{1\text{-}1}$, $P \dfrac{2\text{-}2}{2\text{-}2}$, $M \dfrac{1\text{-}1}{1\text{-}1} = \dfrac{14}{14} = 28$ Lynx

$I \dfrac{2\text{-}2}{1\text{-}1}$, $C \dfrac{0\text{-}0}{0\text{-}0}$, $P \dfrac{3\text{-}3}{2\text{-}2}$, $M \dfrac{3\text{-}3}{3\text{-}3} = \dfrac{16}{12} = 28$ Lepus, Sylvilagus

$I \dfrac{1\text{-}1}{1\text{-}1}$, $C \dfrac{0\text{-}0}{0\text{-}0}$, $P \dfrac{2\text{-}2}{1\text{-}1}$, $M \dfrac{3\text{-}3}{3\text{-}3} = \dfrac{12}{10} = 22$ Marmota, Spermophilus, Glaucomys, Tamias minimus, Sciurus carolinensis

$I \dfrac{1\text{-}1}{1\text{-}1}$, $C \dfrac{0\text{-}0}{0\text{-}0}$, $P \dfrac{1\text{-}1}{1\text{-}1}$, $M \dfrac{3\text{-}3}{3\text{-}3} = \dfrac{10}{10} = 20$ Castor, Erethizon, Geomys, Sciurus niger, Tamiasciurus (typical), Tamias striatus

$I \dfrac{1\text{-}1}{1\text{-}1}$, $C \dfrac{0\text{-}0}{0\text{-}0}$, $P \dfrac{1\text{-}1}{0\text{-}0}$, $M \dfrac{3\text{-}3}{3\text{-}3} = \dfrac{10}{8} = 18$ Zapus

$$I \frac{1\text{-}1}{1\text{-}1}, C \frac{0\text{-}0}{0\text{-}0}, P \frac{0\text{-}0}{0\text{-}0}, M \frac{3\text{-}3}{3\text{-}3} = \frac{8}{8} = 16$$

Clethrionomys, Microtus, Mus, Napaeozapus, Neotoma, Ondatra, Peromyscus, Phenacomys, Rattus, Reithrodontomys, Synaptomys

Technical Names of Organisms Mentioned in the Text

Plants

Alder, *Alnus*
Alfalfa, *Medicago sativa*
American elm, *Ulmus americana*
Apple (common), *Malus pumila*
Arrowhead, *Sagittaria*
Ash, *Fraxinus*
Aspen, *Populus*
Aster, *Aster*
Basswood, *Tilia americana*
Bean, *Vicia*
Bearberry, *Arctostaphylos uva-ursi*
Beech (American), *Fagus grandifolia*
Birch, *Betula*
Blackberry (highbush), *Rubus alleghaniensis*
Black cherry, *Prunus serotina*
Black oak, *Quercus velutina*
Blueberry, *Vaccinium*
Blue grass, *Poa*
Briar, *Rubus*
Bromegrass, *Bromus*
Bulrush, *Scirpus*
Bunchberry, *Cornus canadensis*

Bur oak, *Quercus macrocarpa*
Bur-reed, *Sparganium eurycarpum*
Carrot, *Daucus carota*
Cattail, *Typha*
Cedar
   (red), *Juniperus virginiana*
   (white), *Thuja occidentalis*
Celery, *Spermolepis*
Cherry, *Prunus*
Chickweed (common), *Stellaria media*
Chokecherry, *Prunus virginiana*
Clover, *Trifolium*
Corn, *Zea mays*
Cottonwood, *Populus deltoides*
Crabgrass, *Digitaria*
Cranberry, *Vaccinium macrocarpon*
Currant (wild black), *Ribes americanum*
Daisy (ox-eye), *Chrysanthemum leucanthemum*
Dandelion, *Taraxacum officinale*
Dock (curly), *Rumex crispus*
Dogwood, *Cornus*

Dwarf birch, *Betula pumila*
Elm, *Ulmus*
Fir, *Abies*
Fleabane (annual), *Erigeron annuus*
Foxtail, *Setaria*
Goldenrod (Canada), *Solidago canadensis*
Grape, *Vitis*
Grass, Poaceae
Green ash, *Fraxinus pennsylvanica*
Hawkweed (orange), *Hieracium aurantiacum*
Hazelnut, *Corylus*
Heath, Ericaceae
Hemlock (eastern), *Tsuga canadensis*
Hickory, *Carya*
Honey locust, *Gleditsia triacanthos*
Horsetail, *Equisetum*
Huckleberry, *Gaylussacia*
Larch or Tamarack, *Larix laricina*
Laurel, *Kalmia*
Lettuce, *Lactuca*
Maple, *Acer*
Mayapple, *Podophyllum peltatum*
Mayflower, *Maianthemum canadense*
Milkweed, *Asclepias*
Mulberry (red), *Morus rubra*
Mullein, *Verbascum thapsus*
Oak, *Quercus*
Oats, *Avena sativa*
Pea, *Pisum*
Pine, *Pinus*
Plantain, *Plantago*
Poplar, *Populus*
Potato, *Solanum tuberosa*
Ragweed, *Ambrosia*
Raspberry
  (red), *Rubus idaeus*
  (black), *Rubus occidentalis*

Red maple, *Acer rubrum*
Red oak, *Quercus rubra*
Rhododendron, *Rhododendron*
Rye, *Secale cereale*
Rye (wild), *Elymus*
Rush, *Juncus*
Sedge, *Carex*
Shagbark hickory, *Carya ovata*
Sheep laurel, *Kalmia angustifolia*
Silver maple, *Acer saccharinum*
Soapberry, *Shepherdia canadensis*
Sorrel (wood), *Oxalis*
Soybean, *Glycine max*
Sphagnum, *Sphagnum*
Spruce, *Picea*
Staghorn sumac, *Rhus typhina*
Strawberry, *Fragaria*
Sugar maple, *Acer saccharum*
Sumac, *Rhus*
Sunflower, *Helianthus annuus*
Tamarack, see Larch
Thistle, *Cirsium*
Timothy, *Phleum pratense*
Tomato, *Lycopersicon*
Vetch, *Vicia*
Violet, *Viola*
Water lily
  (white), *Nymphaea*
  (yellow), *Nuphar*
Wheat, *Triticum*
White oak, *Quercus alba*
White pine (eastern), *Pinus strobus*
Willow, *Salix*
Wintergreen, *Gaultheria procumbens*
Yarrow, *Achillea millefolium*

Animals

Ant, Formicidae
Asiatic mouse deer, *Tragulus*
Asiatic shrew mole, *Uropsilus*
Bandicoot, Peramelidae
Barn owl, *Tito alba*
Barred owl, *Strix varia*
Bee (honey), *Apis mellifera*
Beetle, Coleoptera
Birch mouse, *Sicista*
Black-footed cat, *Felis nigripes*
Black rat, *Rattus rattus*
Blue jay, *Cyanocitta cristata*
Blue whale, *Balaenoptera musculus*
Bug (true), Hemiptera
Bullfrog, *Rana catesbeiana*
Bumblebee bat, *Craseonycteris
    thonglongyai*
Caddisfly, Trichoptera
Capybara, *Hydrochaeris hydrochaeris*
Carp, *Cyprinus carpio*
Cattle, *Bos taurus*
Centipede, Chilopoda
Chimney swift, *Chaetura pelagica*
Cheetah, *Acinonyx jubatus*
Chironomid fly, *see* Midge
Cockroach (German), *Blatella
    germanica*
Cooper's hawk, *Accipiter cooperii*
Corn rootworm, *Diabrotica
    undecimpunctata*
Crane, Gruidae
Crayfish, Astacidae
Cricket, Gryllidae
Crow, *Corvus brachyrhynchos*
Darter, *Etheostoma*
Deer tick, *Ixodes dammini*
Dermestid beetle, *Dermestes*
Desman (Russian), *Desmana moschata*
Domestic cat, *see* House cat
Domestic dog, *Canis familiaris*

Dragonfly, Odonata
Earthworm, *Lumbricus*
Fennec fox, *Fennecus zerda*
Fly (true), Diptera
Flying fox, *Pteropus*
Frontal sinus nematode, *Skrjabingylus
    chitwoodorum*
Gerbil, *Meriones unguiculatus*
Giant otter shrew, *Potomogale velox*
Giant panda, *Ailuropoda melanoleuca*
Giraffe, *Giraffa camelopardalis*
Goat (domestic), *Capra hircus*
Golden eagle, *Aquila chrysaetos*
Golden mole, Chrysochloridae
Goshawk, *Accipiter gentilis*
Grackle (common), *Quiscalus quiscula*
Grasshopper, Acrididae
Grasshopper mouse, *Onychomys
    leucogaster*
Great horned owl, *Bubo virginianus*
Grizzly bear, *Ursus arctos*
Gull, (ring-billed), *Larus delawarensis*
Hamster (golden), *Mesocricetus
    auratus*
Harvestman, Opiliones
Harvest mouse (Old World), *Micromys
    minutus*
Hedgehog, Erinaceidae
Hippopotamus, *Hippopotamus
    amphibius*
Horse, *Equus caballus*
House cat, *Felis catus*
Jackass, *Equus asinus*
Jerboa, *Jaculus, Allactaga* (and others)
Kestrel (American), *Falco sparverius*
Kinkajou, *Potos flavus*
Largemouth bass, *Micropterus
    salmoides*
Leafhopper, Cicadellidae
Leech, Hirudinea

Leopard, *Panthera pardus*
Leopard frog, *Rana pipiens*
Lesser panda, *Ailurus fulgens*
Lion, *Panthera leo*
Long-eared owl, *Asio otus*
Mayfly, Ephemeroptera
Mealworm, *Tenebrio molitor*
Meningeal worm, *Parelaphostrongylus tenuis*
Midge, Chironomidae
Millipede, Diplopoda
Minnow, Cyprinidae
Moth, Lepidoptera
Mouse opossum, *Marmosa*
Mussel, Unionidae
Nasal sinus nematode, *Skrjabingylus nasicola*
Peccary, Tayassuidae
Perch (yellow), *Perca flavescens*
Platypus, *Ornithorhynchus anatinus*
Polar bear, *Ursus maritimus*
Polecat, *Mustela putorius*
Porcupine (Old World), Hystricidae
Prairie dog, *Cynomys*
Pudu, *Pudu puda*
Red kangaroo, *Macropus rufus*
Red-tailed hawk, *Buteo jamaicensis*
Rhinoceros, Rhinoceratidae
Roach, Blattidae
Rough-legged hawk, *Buteo lagopus*
Ruffed grouse, *Bonasa umbellus*
Sawfly, *Neodiprion*
Saw-whet owl, *Aegolius acadicus*

Screech owl, *Otus asio*
Sea otter, *Enhydra lutris*
Shrike, *Lanius*
Slender-tailed cloud rat, *Phloemys cumingi*
Slug, Pulmonata
Snail, Pulmonata
Snapping turtle, *Chelydra serpentina*
Solenodon, *Solenodon paradoxus*
Southern elephant seal, *Mirounga leonina*
Sowbug, *Porcellio*
Spider, Araneae
Spiny anteater, Tachyglossidae
Spiny mouse, *Acomys*
Spiny rat, Echimyidae
Spotted bat, *Euderma maculatum*
Springtail, Collembola
Stickleback, Gasterosteidae
Stonefly, Plecoptera
Sucker, Catastomidae
Sunfish, Centrarchidae
Tapir, Tapiridae
Tenrec, Tenrecidae
Tiger, *Panthera tigris*
Trout, *Salmo*
Wasp, Vespidae
Water mouse (Central American), *Rheomys*
Water opossum, *Chironectes minimus*
Whale, Cetacea
Worm, Annelida

Glossary

Altricial. Poorly developed at birth, usually with eyes and ears closed, often lacking hair, and requiring large amounts of parental care; the opposite of precocial.

Angular process. Lowermost projection on posterior part of mandible (fig. 46).

Annulation. Circular ring of scales on a rodent's tail.

Antler. A branched, bony outgrowth of the frontal bone found in members of the deer family; it is shed annually.

Aquatic. Living in water.

Arboreal. Living in trees.

Auditory bulla. A bony capsule housing the middle ear; usually visible as a rounded bulge on the bottom of the skull near the foramen magnum (figs. 46 and 47).

Auditory tube. A bony passage leading into the auditory bulla.

Beam. The main part of an antler.

Bicolored. Having two colors. *See also* distinctly bicolored *and* indistinctly bicolored.

Blastocyst. An early stage of development when an embryo is essentially a hollow ball of cells.

Boreal. Pertaining to the north.

Braincase. Posterior part of the skull that houses the major portion of the brain.

Breech birth. A birth in which the posterior part of a newborn leaves the vagina first.

Brow tine. A branch of an antler that projects forward over the eye.

Cache. To hide food, or the stored food itself.

Calcar. A cartilaginous rod extending from the ankle of a bat and supporting the trailing edge of the tail membrane (*see* fig. 45).

Canid. A member of the dog family (Canidae).

Canine. The tooth immediately behind the incisors, often elongate and conical; in the upper jaw, it is rooted in the maxillary bone and located at the junction of the premaxillary and maxillary bones.

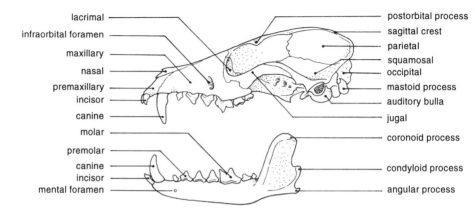

lacrimal — postorbital process
infraorbital foramen — sagittal crest
maxillary — parietal
— squamosal
nasal — occipital
premaxillary — mastoid process
incisor — auditory bulla
canine — jugal
molar — coronoid process
premolar
canine — condyloid process
incisor
mental foramen — angular process

Fig. 46. Lateral view of a red fox skull and mandible with parts labeled

Carnassial pair. Two bladelike teeth, one in the upper and one in the lower jaw, that come together in a scissorslike action to slice meat and crack bones; technically, these teeth are equivalent to the fourth upper premolar and first lower molar of a primitive mammal. Found in many members of the order Carnivora.

Carnivore. Any member of the order Carnivora (dogs, cats, weasels, etc.), regardless of actual diet. In a general sense, any meat-eating animal.

Carnivorous. Meat-eating.

Carrion. Refers to the dead body or flesh of an animal.

Cecum. An outpocketing of the large intestine.

Cervid. A member of the deer family (Cervidae).

Cheek pouch. Paired pockets found in some rodents that are used to carry excess food items, such as seeds; the opening into these pockets may be inside the mouth cavity (internal) or on the side of the face (external).

Cheek teeth. The teeth in back of the canines; the molars and premolars.

Clear-cut. A term used by foresters to refer to an area in which most or all trees have been harvested recently.

Climax. Refers to the mature plant community that results in a particular area over time; the last stage in ecological succession.

Cm. Abbreviation for centimeter, a unit of distance equal to about 0.4 inch; there are about 2.5 cm in 1 inch.

Commensal. In a general sense, it refers to those organisms that live together in close association.

Coprophagy. The habit of eating specially formed fecal pellets.

Cranium. Part of the skull surrounding the brain.

Crepuscular. Active near dawn and dusk.

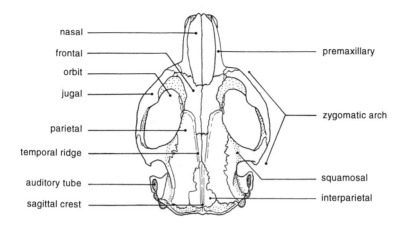

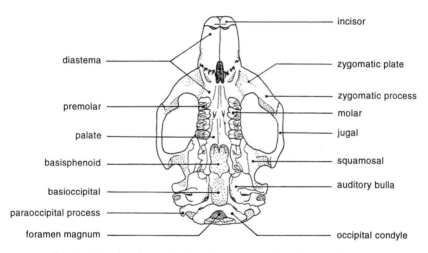

Fig. 47. Dorsal and ventral view of a beaver skull with parts labeled

Cusp. Point or bump on the chewing surface of a tooth.

Daylength. The duration of the day as opposed to night.

Delayed fertilization. A reproductive process in bats. Copulation occurs in late summer, but fertilization (joining of sperm and egg) does not occur until spring, when the female arouses from hibernation; sperm are stored in the uterus throughout winter.

Delayed implantation. A reproductive process commonly found in members of the weasel and bear families in our area. After copulation, the egg is quickly fertilized, develops to the blastocyst stage, and then ceases development for weeks or months; afterward, growth resumes, the embryo attaches to (implants in) the uterine wall, and development continues until birth.

Dentine. Substance that makes up most of a tooth between the inner pulp and outer enamel; often exposed on the chewing surface, particularly in the cheek teeth of herbivorous mammals.

Dewclaw. A vestigial toe, usually small and not contacting the ground.

Diastema. A wide space between teeth; space between the incisors and cheek teeth of rodents and lagomorphs, for example.

Dicot(yledon). A plant with leaves having veins that form a network (rather than parallel to each other), flowers that have parts in multiples of four or five, and seeds having two cotyledons (embryonic leaves).

Didelphid. A member of the opossum family (Didelphidae).

Digitigrade. Walking with only the toes parallel to the ground.

Dipodid. A member of the rodent family Dipodidae, including jumping mice.

Distal. Referring to a part away from the center or base.

Distemper (canine). A viral disease common in members of the dog family and marked by fever and respiratory and nervous problems.

Distinctly bicolored. Refers to a tail in which dorsal and ventral sides abruptly change color (usually dark to light). *See also* indistinctly bicolored.

Diurnal. Active during the day.

Dorsal. Referring to the top or backside of a four-legged animal.

Dorsoventral. From top to bottom.

Ear height. Distance from lowest notch at the base of the ear to the highest point of the pinna, excluding hairs (see fig. 24).

Echolocation. Process by which animals detect and identify objects by emitting high-frequency sounds and interpreting the returning echoes. In our area, used by all bats and some shrews.

Ecosystem. The living (animals and plants) and nonliving (soil, water, energy) components of a particular area.

Enamel. The hard substance that usually forms the exterior of a tooth.

Estivation. A prolonged period of low body temperature and metabolic rate during the warm part of the year.

Estrus. The time when a female is receptive to mating. *See also* postpartum estrus.

Eutherian. A mammal belonging to the Infraclass Eutheria; all living mammals except marsupials and egg-laying monotremes.

Facial mask. A pattern of coloration, usually dark around the eyes, that makes it appear as if the animal is wearing a mask, as in a raccoon.

Fenestrated. Perforated by many irregular openings.

Flank. Side of an animal, especially between the rib cage and hips.

Foramen. An opening into a bone for the passage of nerves, blood vessels, or other soft tissue.

Foramen magnum. The large opening into the skull through which the spinal cord passes (*see* fig. 47).

Forb. A nonwoody, broad-leaved, flowering plant (not a grass or sedge).

Forearm length. The distance from the elbow to the wrist, measured on the folded wing of a bat.

Form. A slight depression in the ground used by a rabbit or hare as a resting site.

Fossorial. Living underground.

Frontal bone. One of a major pair of bones in the forehead area and between the eyes (*see* fig. 47).

Ft. An abbreviation for foot, a unit of distance equal to about 30.5 cm.

G. Abbreviation for gram, a unit of weight equal to about 0.04 ounce; there are about 28 grams in 1 ounce.

Gestation. The period between fertilization of an egg and birth.

Greatest length of skull. Distance from anteriormost tooth or bone to posteriormost projection of skull.

Greatest width of skull. Distance across widest part of skull, usually across zygomatic arches.

Grizzled. Streaked or interspersed with gray.

Grub. The larval form of a beetle.

Guard hair. One of the long hairs covering the shorter and finer hairs that make up the underfur.

Ha. Abbreviation for hectare, a unit of area equalling 10,000 m^2 or about 2.5 acres.

Hallux. The big toe.

Heat. *See* estrus.

Herb. A nonwoody plant.

Herbaceous. Nonwoody.

Herbivore. An animal that primarily eats plants.

Herbivorous. Plant-eating.

Hibernaculum. A place where an animal hibernates.

Hibernation. A prolonged period of low body temperature and metabolic rate during the cold season of the year.

Hindfoot length. Distance measured from the heel to the tip of the claw on the longest toe (*see* fig. 24).

Holarctic. Referring to the northern parts of both the Old World and New World.

Home range. The area in which an animal carries out its daily activities.

Horn. A bony outgrowth of the frontal bone found in many members of the cow family (Bovidae); a horn is covered by a sheath of keratin and is never shed.

Incisor. A front tooth rooted in the premaxillary bone and the corresponding tooth in the lower jaw.

Indistinctly bicolored. Refers to a tail that somewhat gradually changes color from dark on the top to light on the bottom. *See also* distinctly bicolored.

Inflated. Expanded.

Infraorbital foramen. An opening through the zygomatic process of the maxillary bone leading to the orbit (*see* fig. 46).

Insectivore. As used in this book, any member of the order Insectivora (shrews, moles, etc.), regardless of diet. In a general sense, any insect-eating animal.

Insectivorous. Insect-eating.

Interparietal bone. A small bone along the midline at the back of the skull (*see* figs. 33 and 47).

Jugal bone. A bone on each side of the skull forming most of the anterior part of the zygomatic arch (*see* figs. 46 and 47).

Keratin. A tough protein that is a major component of hairs, nails, claws, and other external structures.

Kg. Abbreviation for kilogram, a unit of weight equal to about 2.2 pounds.

kHz. Abbreviation for kilohertz, a unit of frequency, such as the frequency of sound or radio waves; it is equal to 1,000 cycles per second.

Km. Abbreviation for kilometer, a unit of distance equal to about 0.6 mile; there are 1,000 meters in one km.

Lacrimal bone. A small bone that helps form the anterior wall of each orbit; it usually contains a foramen for passage of the tear (lacrimal) duct (*see* fig. 46).

Lactation. The state of producing milk.

Larva. An immature, wormlike form of some insects, such as a caterpillar or maggot.

Laterally compressed. Flattened from side to side.

Lb. Abbreviation for pound, a unit of weight equal to about 0.45 kg.

Least interorbital width. The shortest distance between the orbits measured across the top of the skull.

Leporid. A member of the rabbit family (Leporidae).

Leveret. A young hare.

Loam. A type of soil containing variable amounts of sand, clay, and silt.

M. Abbreviation for meter, a unit of distance equal to about 3.3 feet.

Mandible. The lower jaw.

Masseter muscle. A major muscle involved in chewing.

Mast. Nuts (acorns, walnuts, beechnuts, etc.).

Maxillary bone. One of a pair of large bones contributing to the side of the rostrum and palate; all upper teeth, except incisors, are rooted in the maxillary (*see* figs. 46 and 47).

Maxillary tooth row. The canines, if present, and all cheek teeth.

Medial. Toward the midline.

Melanistic. Having more than the usual amount of the pigment melanin, resulting in a darker-than-normal animal.

Mental foramen. A small opening on the anterior part of the dentary bone (mandible) (*see* fig. 46).

Metabolic rate. The rate at which an animal uses energy.

Metabolic water. Water formed mainly as a by-product of the chemical breakdown of food molecules within the body.

Mist net. A very fine net made of nylon and used to capture flying birds or bats.

Mm. Abbreviation for millimeter, a unit of distance equal to about 0.04 inch; there are 10 mm in one centimeter and 1,000 mm in one meter.

Molar. A posterior cheek tooth, usually with multiple roots. A mammal has only one set of molars in its life; there are no molars among the deciduous (milk) teeth.

Monestrous. Coming into heat (estrus) only once each year.

Monocot(yledon). A flowering plant with parallel veins in its leaves, flowers that have parts in multiples of three, and seeds having one cotyledon (embryonic leaf).

Morphology. The features that contribute to the form and structure of an animal or its parts.

Murid. A member of a rodent family (Muridae) that includes mice, rats, voles, and lemmings, among others.

Mustelid. A member of the weasel family (Mustelidae).

Nape. The back of the neck.

Nasal bone. One of a pair of bones partly covering the nasal cavity (see figs. 46 and 47).

Niche. The role of an organism in its community.

Nocturnal. Active at night.

Occipital area. The area surrounding the foramen magnum.

Occlusal. An adjective referring to the chewing surface of a tooth.

Olfaction. The sense of smell.

Omnivore. An animal that eats many types of foods, including plants and animals.

Opposable. Capable of being placed opposite something. As an example, the human thumb can be placed opposite other fingers to grasp an object; the thumb is opposable.

Orbit. The bony socket of the eye.

Oz. Abbreviation for ounce, a unit of weight equal to about 0.03 kilogram.

Palate. The roof of the mouth.

Palmate. Shaped somewhat like a hand with fingers spread.

Parietal bone. One of a pair of bones that forms much of the top of the skull behind the frontal bones (see figs. 46 and 47).

Parturition. The act of giving birth.

Patagium. The fleshy membrane that allows a flying squirrel to glide or a bat to fly.

Pelage. A general term referring to the hairy coat of a mammal.

Photoperiod. *See* daylength.

Pinna. The external ear.

Placental. *See* eutherian.

Plantar tubercle. A small nodule or bump on the sole of the foot.

Pollex. The thumb.

Polyestrous. Coming into heat (estrus) more than once each year.

Postglenoid length of skull. Distance from the glenoid cavity (depression in skull where the lower jaw meets the upper jaw) to the most posterior part of skull.

Postorbital process. A projection off the frontal bone above and behind the orbit (*see* fig. 46).

Postpartum estrus. A period of sexual receptivity just after a female gives birth.

Precocial. Well developed at birth, fully furred, and able to see, hear, and walk to a certain degree; the opposite of altricial.

Prehensile. Referring to tails that can be used for grasping tree branches or other objects.

Premaxillary bone. One of a pair of bones at the anterior end of the rostrum (*see* fig. 46); the upper incisors are rooted in the premaxillary bones.

Premolar. A cheek tooth between the molars and canine.

Presphenoid. An unpaired bone with a complex shape; it may be seen along the midline posterior to the palate (*see* fig. 36) or as part of the orbit wall.

Process. Any projection off the main part of a bone.

Promiscuous. Having more than one mate in a single season.

Proximal. Referring to a part close to the base or center.

Pupa. An inactive stage in the development of an insect between the larval and adult stage; for example, a butterfly pupa is hidden inside a cocoon.

Rabies. A viral disease that attacks the central nervous system.

Raptor. Predatory birds, such as hawks and owls.

Recurved. Curving backward.

Reentrant angle. An infolding of the enamel bordering a cheek tooth, especially in voles and related rodents (*see* fig. 35).

Retina. The layer of light-sensitive neurons inside the eye responsible for sight.

Retractile. A word describing the ability of some mammals, especially cats, to pull the claws up and back so that they do not contact the surface of the ground.

Rhizome. A rootlike stem often located underground.

Riparian. Associated with streams or rivers.

Rostrum. The part of the skull that is anterior to the orbits.

Runway. Tunnellike passage through grassy vegetation commonly built by voles, mice, and shrews.

Rut. The annual period of sexual activity in male deer characterized by male-male aggression and by mating.

Sagittal crest. Ridge of bone along the dorsal midline of the braincase used for muscle attachment (see figs. 46 and 47); paired temporal ridges often come together to form a sagittal crest.

Saltatorial. Referring to movement by hopping.

Sarcoptic mange. Skin disease that results in loss of hair and is caused by mites in the genus *Sarcoptes.*

Scatter hoard. To hide food items individually as opposed to gathering them in piles.

Sciurid. A member of the squirrel family (Sciuridae).

Scrotum. The pouch of skin that encloses the testes.

Soricid. A member of the shrew family (Soricidae).

Spine. A pointed projection (process) coming off the main part of a bone.

Subnivean. Under the snow.

Supraorbital process. A projection from the frontal bone above the orbit; it is well developed in rabbits and hares (see fig. 33).

Suture. A joint between two bones commonly found in the skull.

Swarming. A behavior of bats, primarily *Myotis* and *Pipistrellus,* that occurs in late summer and involves many bats visiting a mine or cave for brief periods at night; mating often takes place during swarming.

Tail length. Distance from the base of the tail to the tip of the last vertebra (see fig. 24).

Tail membrane. A flap of skin stretching from one leg to the other and enclosing the tail vertebrae of a bat.

Talpid. A member of the mole family (Talpidae).

Temporal ridge. Ridge of bone on top of the skull behind the orbit; ridges from opposite sides converge at the rear of the skull and may join to form a sagittal crest (see fig. 47).

Terrestrial. Living primarily on the ground.

Territory. A space defended by an animal against others of its kind.

Tine. A branch of an antler.

Torpor. A state of reduced body temperature and metabolic rate.

Total length. Distance from the nose to the tip of the last tail vertebra (see fig. 24).

Tragus. A fleshy projection at the base of the ear opening, particularly in a bat (see fig. 45).

Tricolored. Having three colors; may refer to an individual hair or to the overall appearance of an animal's body.

Tuber. A thickened underground stem (rhizome) that contains stored food; a potato is a tuber.

Tularemia. A bacterial disease of rabbits, hares, and rodents that may be transmitted to humans who contact an infected carcass.

Underfur. The hairs of a mammal's coat that provide most of the insulation; hairs in the underfur are typically shorter and thinner than guard hairs.

Unicuspid. Generally, any tooth with a single cusp. Specifically, the row of simple teeth in a shrew behind the first incisor and in front of the first complex cheek tooth (*see* fig. 30).

Ursid. A member of the bear family (Ursidae).

Ventral. Referring to the lower or belly side of a four-legged animal.

Vertebra. One of the chain of bones forming the "back bone" and tail.

Vespertilionid. A member of the plain-nosed bat family (Vespertilionidae).

Vibrissae. Long, stiff hairs located in the facial region; the "whiskers."

Vole. A type of small rodent in the genera *Clethrionomys, Microtus,* or *Phenacomys.*

Zygomatic arch. A narrow strip of bone that begins at the anterior edge of the orbit and extends back to the braincase (*see* fig. 47); it often consists of part of the maxillary bone, the jugal bone, and part of the squamosal bone.

Zygomatic plate. Wide, flattened area at the anterior end of the zygomatic arch (*see* fig. 47); the plate is actually the basal part of the zygomatic process of the maxillary bone.

References

Albert, D. A., S. R. Denton, and B. V. Barnes. 1986. *Regional landscape ecosystems of Michigan*. Ann Arbor: School Nat. Res., Univ. Michigan.

Anthony, R. G., D. A. Simpson, and G. M. Kelly. 1986. Dynamics of pine vole populations in two Pennsylvania orchards. *Amer. Midl. Nat.* 116:108–17.

Arthur, S. M., and W. B. Krohn. 1991. Activity patterns, movements, and reproductive ecology of fishers in southcentral Maine. *J. Mamm.* 72:379–85.

Baird, D. D., R. M. Timm, and G. E. Nordquist. 1983. Reproduction in the arctic shrew, *Sorex arcticus*. *J. Mamm.* 64:298–301.

Baker, R. H. 1983. *Michigan mammals*. East Lansing: Michigan State Univ. Press.

Baker, R. J., and S. L. Williams. 1972. A live trap for pocket gophers. *J. Wildl. Manage.* 36:1320–22.

Banfield, A. W. F. 1974. *The mammals of Canada*. Toronto: Univ. Toronto Press.

Barbour, R. W., and W. H. Davis. 1969. *Bats of America*. Louisville: Univ. Press Kentucky.

Barclay, R. M. R. 1989. The effect of reproductive condition on the foraging behavior of female hoary bats, *Lasiurus cinereus*. *Behav. Ecol. Sociobiol.* 24:31–37.

Barclay, R. M. R., P. A. Faure, and D. R. Farr. 1988. Roosting behavior and roost selection by migrating silver-haired bats (*Lasionycteris noctivagans*). *J. Mamm.* 69:821–25.

Beckel, A. L. 1991. Wrestling play in adult river otters. *J. Mamm.* 72:386–90.

Bekoff, M. 1977. *Canis latrans*. *Mamm. Species* 79:1–9.

Beneski, J. T., Jr., and D. W. Stinson. 1987. *Sorex palustris*. *Mamm. Species* 296:1–6.

Beyer, D. E., Jr. 1987. Population and habitat management of elk in Michigan. Ph.D. diss., Michigan State Univ.

Bittner, S. L., and O. J. Rongstad. 1982. Snowshoe hare and allies. In *Wild mammals of North America*, ed. J. A. Chapman and G. A. Feldhamer, 146–63. Baltimore: Johns Hopkins Univ. Press.

Boutin, S. 1992. Predation and moose population dynamics: a critique. *J. Wildl. Manage.* 56:116–27.

Braun, E. L. 1964. *Deciduous forests of eastern North America.* New York: Hafner.

Bronson, F. H. 1979. The reproductive ecology of the house mouse. *Quart. Rev. Biol.* 54:265–99.

Buckner, C. H. 1970. Direct observations of shrew predation on insects and fish. *Blue Jay* 28:171–72.

Chapman, J. A., K. L. Cramer, N. J. Dippenaar, and T. J. Robinson. 1992. Systematics and biogeography of the New England cottontail, *Sylvilagus transitionalis* (Bangs, 1895), with the description of a new species from the Appalachian Mountains. *Proc. Biol. Soc. Washington* 105:841–66.

Chapman, J. A., and G. A. Feldhamer, eds. 1982. *Wild mammals of North America.* Baltimore: Johns Hopkins Univ. Press.

Chapman, J. A., J. G. Hockman, and W. R. Edwards. 1982. Cottontails. In *Wild mammals of North America,* ed. J. A. Chapman and G. A. Feldhamer, 83–145. Baltimore: Johns Hopkins Univ. Press.

Chapman, J. A., J. G. Hockman, and M. M. Ojeda C. 1980. *Sylvilagus floridanus. Mamm. Species* 136:1–8.

Chapman, L. J., and D. F. Putnam. 1973. *The physiography of southern Ontario.* 2d ed. Toronto: Univ. Toronto Press.

Choate, J. R., ed. 1987. Acceptable field methods in mammalogy: Preliminary guidelines approved by the American Society of Mammalogists. *J. Mamm.* 68 (Suppl.): 1–18.

Choromanski-Norris, J., E. K. Fritzell, and A. B. Sargeant. 1986. Seasonal activity cycle and weight changes of the Franklin's ground squirrel. *Amer. Midl. Nat.* 116:101–7.

————. 1989. Movements and habitat use of Franklin's ground squirrels in duck-nesting habitat. *J. Wildl. Manage.* 53:324–31.

Christian, D. P., and J. M. Daniels. 1985. Distributional records of rock voles, *Microtus chrotorrhinus,* in northeastern Minnesota. *Can. Field-Nat.* 99:356–59.

Clark, T. W., E. Anderson, C. Douglas, and M. Strickland. 1987. *Martes americana. Mamm. Species* 289:1–8.

Clark, W. R., J. J. Hasbrouck, J. M. Kienzler, and T. F. Glueck. 1989. Vital statistics and harvest of an Iowa raccoon population. *J. Wildl. Manage.* 53:982–90.

Clayton, J. S., W. A. Ehrlich, D. B. Cann, J. H. Day, and I. B. Marshall. 1977. *Soils of Canada. Vol. 1.* Ottawa: Research Branch, Canada Dept. Agr.

Clough, G. C. 1963. Biology of the Arctic shrew, *Sorex arcticus. Amer. Midl. Nat.* 69:69–81.

Clough, G. C., and J. J. Albright. 1987. Occurrence of the northern bog lemming,

Synaptomys borealis, in the northeastern United States. *Can. Field-Nat.* 101:611–13.

Constantine, D. G. 1988. Health precautions for bat researchers. In *Ecological and behavioral methods for the study of bats,* ed. T. H. Kunz, 491–528. Washington, D. C.: Smithsonian Inst. Press.

Currier, M. J. P. 1983. *Felis concolor. Mamm. Species* 200:1–7.

Darby, W. R., and W. O. Pruitt, Jr. 1984. Habitat use, movements and grouping behaviour of woodland caribou, *Rangifer tarandus,* in southeastern Manitoba. *Can. Field-Nat.* 98:184–210.

Dean, P. B., and A. de Vos. 1965. The spread and status of the European hare, *Lepus europaeus hybridus* (Desmarest) in North America. *Can. Field-Nat.* 79:38–48.

DeBlase, A. F., and R. E. Martin. 1981. *A manual of mammalogy with keys to families of the world.* 2d ed. Dubuque: W. C. Brown.

Dolan, P. G., and D. C. Carter. 1977. *Glaucomys volans. Mamm. Species* 78:1–6.

Dorr, J. A., Jr., and D. F. Eschman. 1970. *Geology of Michigan.* Ann Arbor: Univ. Michigan Press.

Dyck, A. P., and R. A. MacArthur. 1993. Seasonal variation in the microclimate and gas composition of beaver lodges in a boreal environment. *J. Mamm.* 74:180–88.

Eichenlaub, V. L. 1979. *Weather and climate of the Great Lakes region.* Notre Dame, Ind.: Univ. Notre Dame Press.

Faure, P. A., J. H. Fullard, and J. W. Dawson. 1993. The gleaning attacks of the northern long-eared bat, *Myotis septentrionalis,* are relatively inaudible to moths. *J. Exp. Biol.* 178:173–89.

Fenton, M. B. 1972. Distribution and overwintering of *Myotis leibii* and *Eptesicus fuscus. Life Sci. Contr.,* Roy. Ontario Mus., 77:1–18.

Fenton, M. B., and R. M. R. Barclay. 1980. *Myotis lucifugus. Mamm. Species* 142:1–8.

Fitch, H. S. 1950. A new style live-trap for small mammals. *J. Mamm.* 31:364–65.

Fitch, J. H., and K. A. Shump, Jr. 1979. *Myotis keenii. Mamm. Species* 121:1–3.

Foltz, D. W., and P. L. Schwagmeyer. 1989. Sperm competition in the thirteen-lined ground squirrel: Differential fertilization success under field conditions. *Amer. Nat.* 133:257–65.

Forbes, R. B. 1966. Studies of the biology of Minnesotan chipmunks. *Amer. Midl. Nat.* 76:290–308.

Forman, G. L., and C. J. Phillips. 1988. Preparation and fixation of tissues for histological, histochemical, immunohistochemical, and electron microscopical studies. In *Ecological and behavioral methods for the study of bats,* ed. T. H. Kunz, 405–24. Washington, D. C.: Smithsonian Inst. Press.

Forsyth, D. J. 1976. A field study of growth and development of nestling masked shrews (*Sorex cinereus*). *J. Mamm.* 57:708–21.

Franzmann, A. W. 1981. *Alces alces. Mamm. Species* 154:1–7.

French, T. W. 1980a. Natural history of the southeastern shrew, *Sorex longirostris* Bachman. *Amer. Midl. Nat.* 104:13–31.

———. 1980b. *Sorex longirostris. Mamm. Species* 143:1–3.

———. 1984. Dietary overlap of *Sorex longirostris* and *S. cinereus* in hardwood floodplain habitats in Vigo County, Indiana. *Amer. Midl. Nat.* 111:41–46.

Fridell, R. A., and J. A. Litvaitis. 1991. Influence of resource distribution and abundance on home-range characteristics of southern flying squirrels. *Can. J. Zool.* 69:2589–93.

Fritts, S. H., and L. D. Mech. 1981. Dynamics, movements, and feeding ecology of a newly protected wolf population in northwestern Minnesota. *Wildl. Monogr.* 80:1–79.

Fritts, S. H., W. J. Paul, L. D. Mech, and D. P. Scott. 1992. Trends and management of wolf-livestock conflicts in Minnesota. *United States Fish Wildl. Serv., Resource Publ.* 181:1–27.

Fritzell, E. K., and K. J. Haroldson. 1982. *Urocyon cinereoargenteus. Mamm. Species* 189:1–8.

Fujita, M. S., and T. H. Kunz. 1984. *Pipistrellus subflavus. Mamm. Species* 228:1–6.

Gardner, A. L. 1982. Virginia opossum. In *Wild mammals of North America,* ed. J. A. Chapman and G. A. Feldhamer, 3–36. Baltimore: Johns Hopkins Univ. Press.

Gardner, J. E., J. D. Garner, and J. E. Hofmann. 1991. Summer roost selection and roosting behavior of *Myotis sodalis* (Indiana bat) in Illinois. Illinois Nat. Hist. Surv., Champaign. Photocopy.

Gates, D. M., C. H. D. Clarke, and J. T. Harris. 1983. Wildlife in a changing environment. In *The Great Lakes forest: An environmental and social history,* ed. S. L. Flader, 52–82. Minneapolis: Univ. Minnesota Press.

George, S. B., J. R. Choate, and H. H. Genoways. 1986. *Blarina brevicauda. Mamm. Species* 261:1–9.

Gerson, H. B. 1988. Cougar, *Felis concolor,* sightings in Ontario. *Can. Field-Nat.* 102:419–24.

Getz, L. L., B. McGuire, T. Pizzuto, J. E. Hofmann, and B. Frase. 1993. Social organization of the prairie vole (*Microtus ochrogaster*). *J. Mamm.* 74:44–58.

Gilbert, A. N. 1986. Mammary number and litter size in Rodentia: The "one-half rule." *Proc. Natl. Acad. Sci.* 83:4828–30.

Glass, B. P. 1981. *Key to the skulls of North American mammals.* 2d ed. Stillwater: Dept. Zool., Univ. Oklahoma.

Gottschang, J. L. 1981. *A guide to the mammals of Ohio*. Columbus: Ohio State Univ. Press.

Gould, E., W. McShea, and T. Grand. 1993. Function of the star in the star-nosed mole, *Condylura cristata*. *J. Mamm.* 74:108–16.

Hall, E. R. 1981. *The mammals of North America*. 2d ed. New York: John Wiley.

Hallett, J. G. 1978. *Parascalops breweri*. *Mamm. Species* 98:1–4.

Hamilton, W. J., Jr., and J. O. Whitaker, Jr. 1979. *Mammals of the eastern United States*. Ithaca, N. Y.: Cornell Univ. Press.

Handley, C. O., Jr. 1988. Specimen preparation. In *Ecological and behavioral methods for the study of bats*, ed. T. H. Kunz, 437–58. Washington, D.C.: Smithsonian Inst. Press.

Harlow, H. J. 1981. Torpor and other physiological adaptations of the badger (*Taxidea taxus*) to cold environments. *Physiol. Zool.* 54:267–75.

Hayes, J. P., and M. E. Richmond. 1993. Clinal variation and morphology of woodrats (*Neotoma*) of the eastern United States. *J. Mamm.* 74:204–16.

Hazard, E. B. 1982. *The mammals of Minnesota*. Minneapolis: Univ. Minnesota Press.

Heinrich, B. 1992. Maple sugaring by red squirrels. *J. Mamm.* 73:51–54.

Henry, J. D. 1986. *Red fox: The catlike canine*. Washington, D. C.: Smithsonian Inst. Press.

Hickey, M. B. C., and M. B. Fenton. 1990. Foraging by red bats (*Lasiurus borealis*): Do intraspecific chases mean territoriality? *Can. J. Zool.* 68:2477–82.

Hitchcock, H. B., R. Keen, and A. Kurta. 1984. Survival rates of *Myotis leibii* and *Eptesicus fuscus* in southeastern Ontario. *J. Mamm.* 65:126–30.

Hoffmeister, D. F. 1989. *Mammals of Illinois*. Urbana: Univ. Illinois Press.

Horner, M. A., and R. A Powell. 1990. Internal structure of home ranges of black bears and analyses of home-range overlap. *J. Mamm.* 71:402–10.

Hossler, R. J., J. B. McAninch, and J. D. Harder. 1994. Maternal denning behavior and survival of juveniles in opossums in southeastern New York. *J. Mamm.* 75:60–70.

Hough, J. L. 1958. *Geology of the Great Lakes*. Urbana: Univ. Illinois Press.

Howard, W. E., and R. E. Marsh. 1982. Spotted and hog-nosed skunks. In *Wild mammals of North America*, ed. J. A. Chapman and G. A. Feldhamer, 664–73. Baltimore: Johns Hopkins Univ. Press.

Hoyle, J. A., and R. Boonstra. 1986. Life history traits of the meadow jumping mouse, *Zapus hudsonius*, in southern Ontario. *Can. Field-Nat.* 100:537–44.

Irvin, A. D., and J. E. Cooper. 1972. Possible health hazards with the collection and handling of post-mortem zoological material. *Mamm. Rev.* 2:43–54.

Iverson, S. L., and B. N. Turner. 1972. Natural history of a Manitoba population of Franklin's ground squirrels. *Can. Field-Nat.* 86:145–49.

Jackson, H. H. T. 1961. *Mammals of Wisconsin*. Madison: Univ. Wisconsin Press.

Jackson, W. B. 1982. Norway rat and allies. In *Wild mammals of North America*, ed. J. A. Chapman and G. A. Feldhamer, 1077–88. Baltimore: Johns Hopkins Univ. Press.

Jenkins, S. H., and P. E. Busher. 1979. *Castor canadensis*. *Mamm. Species* 120:1–8.

Jones, E. M., and R. D. Owen. 1987. Fluid preservation of specimens. In *Mammal collection management*, ed. H. H. Genoways, C. Jones, and O. L. Rossolimo, 51–64. Lubbock: Texas Tech Press.

Jones, J. K., Jr., and E. C. Birney. 1988. *Handbook of mammals of the north-central states*. Minneapolis: Univ. Minnesota Press.

Jones, J. K., Jr., R. S. Hoffmann, D. W. Rice, C. Jones, R. J. Baker, and M. D. Engstrom. 1992. Revised checklist of North American mammals north of Mexico, 1991. *The Museum, Texas Tech Univ., Occ. Pap.* 146.

Jones, J. K., Jr., and R. W. Manning. 1992. *Illustrated key to skulls of genera of North American land mammals*. Lubbock: Texas Tech Press.

Keith, L. B., and S. E. M. Bloomer. 1993. Differential mortality of sympatric snowshoe hares and cottontail rabbits in central Wisconsin. *Can. J. Zool.* 71:1694–97.

Kent, G. C. 1992. *Comparative anatomy of the vertebrates*. 7th ed. St. Louis: Mosby Yearbook.

King, C. M. 1983. *Mustela erminea*. *Mamm. Species* 195:1–8.

———. 1990. *The natural history of weasels and stoats*. Ithaca, N. Y.: Comstock Publishing Assoc.

King, J. A., ed. 1968. Biology of *Peromyscus*. *Amer. Soc. Mamm., Spec. Publ.*, 2.

Kirkland, G. L., Jr. 1981. *Sorex dispar* and *Sorex gaspensis*. *Mamm. Species* 155:1–4.

Kirkland, G. L., Jr., and F. J. Jannett, Jr. 1982. *Microtus chrotorrhinus*. *Mamm. Species* 180:1–5.

Kivett, V. K., and O. B. Mock. 1980. Reproductive behavior in the least shrew (*Cryptotis parva*) with special reference to the aural glandular region of the female. *Amer. Midl. Nat.* 103:339–45.

Klein, H. G. 1960. Ecological relationships of *Peromyscus leucopus noveboracensis* and *Peromyscus maniculatus gracilis* in central New York. *Ecol. Monogr.* 30:387–487.

Knudsen, J. 1972. *Collecting and preserving plants and animals*. New York: Harper and Row.

Koprowski, J. L. 1994a. *Sciurus carolinensis*. *Mamm. Species*. In press.

———. 1994b. *Sciurus niger*. *Mamm. Species*. In press.

Kunz, T. H. 1982. *Lasionycteris noctivagans*. *Mamm. Species* 172:1–5.

Kunz, T. H., and A. Kurta. 1988. Capture methods and holding devices. In *Eco-

logical and behavioral methods for the study of bats, ed. T. H. Kunz, 1–30. Washington, D. C.: Smithsonian Inst. Press.

Kurta, A., and R. H. Baker. 1990. *Eptesicus fuscus. Mamm. Species* 356:1–10.

Kurta, A., G. P. Bell, K. A. Nagy, and T. H. Kunz. 1989. Energetics of pregnancy and lactation in free-ranging little brown bats (*Myotis lucifugus*). *Physiol. Zool.* 62:804–18.

Kurta, A., D. King, J. A. Teramino, J. M. Stribley, and K. J. Williams. 1993. Summer roosts of the endangered Indiana bat (*Myotis sodalis*) on the northern edge of its range. *Amer. Midl. Nat.* 129:132–38.

Kurta, A., and M. E. Stewart. 1990. Parturition in the silver-haired bat, *Lasionycteris noctivagans,* with a description of the neonates. *Can. Field-Nat.* 104:598–600.

Lackey, J. A., D. G. Huckaby, and B. G. Ormiston. 1985. *Peromyscus leucopus. Mamm. Species* 247:1–10.

Lair, H. 1985. Length of gestation in the red squirrel, *Tamiasciurus hudsonicus. J. Mamm.* 66:809–10.

Lavigueur, L., and C. Barrette. 1992. Suckling, weaning, and growth in captive woodland caribou. *Can. J. Zool.* 70:1753–66.

Lee, D. S., and J. B. Funderburg. 1982. Marmots. In *Wild mammals of North America,* ed. J. A. Chapman and G. A. Feldhamer, 176–229. Baltimore: Johns Hopkins Univ. Press.

Lim, B. K. 1987. *Lepus townsendii. Mamm. Species* 288:1–6.

Linscombe, G., N. Kinler, and R. J. Aulerich. 1982. Mink. In *Wild mammals of North America,* ed. J. A. Chapman and G. A. Feldhamer, 629–43. Baltimore: Johns Hopkins Univ. Press.

Linzey, A. V. 1983. *Synaptomys cooperi. Mamm. Species* 210:1–5.

Lomolino, M. V. 1993. Winter filtering, immigrant selection and species composition of insular mammals of Lake Huron. *Ecography* 16:24–30.

———. 1994. Species richness of mammals inhabiting nearshore archipelagoes: area, isolation, and immigration filters. *J. Mamm.* 75:39–49.

Long, C. A. 1973. *Taxidea taxus. Mamm. Species* 26:1–4.

———. 1974. *Microsorex hoyi* and *Microsorex thompsoni. Mamm. Species* 33:1–4.

Lotze, L.-H., and S. Anderson. 1979. *Procyon lotor. Mamm. Species* 119:1–8.

Madison, D. M., and W. J. McShea. 1987. Seasonal changes in reproductive tolerance, spacing, and social organization in meadow voles: A microtine model. *Amer. Zool.* 27:899–908.

Maldonado, J., and G. L. Kirkland, Jr. 1986. Relationship between cranial damage attributable to *Skrjabingylus* (Nematoda) and braincase capacity in the striped skunk. *Can. J. Zool.* 64:2004–7.

Marinelli, L., and F. Messier. 1993. Space use and the social system of muskrats. *Can. J. Zool.* 71:869–75.

Martin, I. G. 1981. Venom of the short-tailed shrew (*Blarina brevicauda*) as an immobilizing agent. *J. Mamm.* 62:189–92.

McAllister, J. A., and R. S. Hoffman. 1988. *Phenacomys intermedius. Mamm. Species* 305:1–8.

McCord, C. M., and J. E. Cardoza. 1982. Bobcat and lynx. In *Wild mammals of North America,* ed. J. A. Chapman and G. A. Feldhamer, 728–66. Baltimore: Johns Hopkins Univ. Press.

McManus, J. J. 1974. *Didelphis virginiana. Mamm. Species* 40:1–6.

Meagher, M. 1986. *Bison bison. Mamm. Species* 266:1–8.

Mech, L. D. 1974. *Canis lupus. Mamm. Species* 37:1–6.

Melquist, W. E., and A. E. Hornocker. 1983. Ecology of river otters in west central Idaho. *Wildl. Monogr.* 80:1–160.

Merritt, J. F. 1981. *Clethrionomys gapperi. Mamm. Species* 146:1–9.

———. 1987. *Guide to the mammals of Pennsylvania.* Pittsburgh: Univ. Pittsburgh Press.

Merritt, J. F., and J. M. Merritt. 1978. Population ecology and energy relationships of *Clethrionomys gapperi* in a Colorado subalpine forest. *J. Mamm.* 59:576–98.

Messick, J. P., and M. G. Hornocker. 1981. Ecology of the badger in southwestern Idaho. *Wildl. Monogr.* 76:1–53.

Mikesic, D. G., and L. C. Drickamer. 1992. Factors affecting home-range size in house mice (*Mus musculus domesticus*) living in outdoor enclosures. *Amer. Midl. Nat.* 127:31–40.

Miller, F. L. 1982. Caribou. In *Wild mammals of North America,* ed. J. A. Chapman and G. A. Feldhamer, 923–59. Baltimore: Johns Hopkins Univ. Press.

Mumford, R. E., and J. O. Whitaker, Jr. 1982. *Mammals of Indiana.* Bloomington: Indiana Univ. Press.

Murie, J. O., and G. R. Michener, eds. 1984. *The biology of ground-dwelling squirrels.* Lincoln: Univ. Nebraska Press.

Nagorsen, D. W. 1987. Summer and winter food caches of the heather vole, *Phenacomys intermedius,* in Quetico Provincial Park, Ontario. *Can. Field-Nat.* 101:82–85.

Nagorsen, D. W., and R. L. Peterson. 1980. *Mammal collector's guide.* Toronto: Roy. Ontario Mus.

Nowak, R. M. 1991. *Walker's mammals of the world.* 2 vols. Baltimore: Johns Hopkins Univ. Press.

Nudds, T. D. 1990. Retroductive logic in retrospect: The ecological effects of meningeal worms. *J. Wildl. Manage.* 54:396–402.

Oftedal, O. T. 1984. Milk composition, milk yield and energy output at peak lactation: A comparative review. *Symp. Zool. Soc. Lond.* 51:33–85.

Oliveras, D., and M. Novak. 1986. A comparison of paternal behaviour in the meadow vole (*Microtus pennsylvanicus*), the pine vole (*M. pinetorum*), and the prairie vole (*M. ochrogaster*). *Anim. Behav.* 34:519–26.

Owen, J. G. 1984. *Sorex fumeus. Mamm. Species* 215:1–8.

Ozoga, J. J., and C. J. Phillips. 1964. Mammals of Beaver Island, Michigan. *Michigan State Univ., Publ. Mus., Biol. Ser.* no. 2:305–48.

Pelton, M. R. 1982. Black bear. In *Wild mammals of North America.*, ed. J. A. Chapman and G. A. Feldhamer, 504–14. Baltimore: Johns Hopkins Univ. Press.

Petersen, K. E., and T. L. Yates. 1980. *Condylura cristata. Mamm. Species* 129:1–4.

Peterson, R. L. 1966. *The mammals of eastern Canada.* Toronto: Oxford Univ. Press.

Poole, E. L. 1940. A life history sketch of the Allegheny woodrat. *J. Mamm.* 21:249–70.

Powell, R. A. 1981. *Martes pennanti. Mamm. Species* 159:1–6.

———. 1982. *The fisher: Life history, ecology, and behavior.* Minneapolis: Univ. Minnesota Press.

Ranta, W. B., H. G. Merriam, and J. F. Wegner. 1982. Winter habitat use by wapiti, *Cervus elaphus,* in Ontario woodlands. *Can. Field-Nat.* 96:421–30.

Reich, L. M. 1981. *Microtus pennsylvanicus. Mamm. Species* 159:1–8.

Reimers, E. 1993. Antlerless females among reindeer and caribou. *Can. J. Zool.* 71:1319–25.

Ropek, R. M., and R. K. Neely. 1993. Mercury levels in Michigan river otters, *Lutra canadensis. J. Freshwater Ecol.* 8:141–47.

Rose, R. K. 1973. A small mammal live trap. *Trans. Kansas Acad. Sci.* 76:14–17.

Roze, U. 1989. *The North American porcupine.* Washington, D. C.: Smithsonian Inst. Press.

Ryan, J. M. 1986. Dietary overlap in sympatric populations of pygmy shrews, *Sorex hoyi,* and masked shrews, *Sorex cinereus,* in Michigan. *Can. Field-Nat.* 100:225–28.

Scharf, W. C. 1973. Birds and land vertebrates of South Manitou Island. *Jack-Pine Warbler* 51:3–19.

Scharf, W. C., and M. L. Jorae. 1980. Birds and land vertebrates of North Manitou Island. *Jack-Pine Warbler* 58:4–15.

Schmitz, O. J., and G. B. Kolenosky. 1985. Wolves and coyotes in Ontario: Morphological relationships and origins. *Can. J. Zool.* 63:1130–37.

Schneider, E. 1990. Hares and rabbits. In *Grzimek's encyclopedia of mammals,* ed. S. B. Parker, 4:254–313. New York: McGraw-Hill.

Schug, M. D., S. H. Vessey, and A. I. Korytko. 1991. Longevity and survival in a population of white-footed mice (*Peromyscus leucopus*). *J. Mamm.* 72:360–66.

Sheffield, S. R., and C. M. King. 1994. *Mustela nivalis. Mamm. Species* 454:1–9.

Sheldon, J. W. 1991. *Wild dogs: The natural history of the nondomestic* Canidae. New York: Academic Press.

Shemanchuk, J. A., and H. J. Bergen. 1968. The gen trap, a simple, humane trap for Richardson's ground squirrels, *Citellus richardsonii* (Sabine). *J. Mamm.* 49:553–55.

Sherburne, S. S. 1993. Squirrel middens influence marten (*Martes americana*) use of subnivean access points. *Amer. Midl. Nat.* 129:204–7.

Sheridan, P. 1978. Formax preserved birds. *Amer. Biol. Teacher* 40:21–22, 35.

Shump, K. A., Jr., and A. U. Shump. 1982a. *Lasiurus borealis. Mamm. Species* 183:1–6.

———. 1982b. *Lasiurus cinereus. Mamm. Species* 185:1–5.

Simms, S. A. 1979a. North American weasels: Resource utilization and distribution. *Can. J. Zool.* 57:504–20.

———. 1979b. Studies of an ermine population in southern Ontario. *Can. J. Zool.* 57:824–32.

Sinclair, A. R. E., et al. 1993. Can the solar cycle and climate synchronize the snowshoe hare cycle in Canada? Evidence from tree rings and ice cores. *Amer. Nat.* 141:173–98.

Smith, W. P. 1991. *Odocoileus virginianus. Mamm. Species* 388:1–13.

Smolen, M. J. 1981. *Microtus pinetorum. Mamm. Species* 147:1–7.

Snyder, D. P. 1982. *Tamias striatus. Mamm. Species* 168:1–8.

Sommer, H. G., and S. Anderson. 1974. Cleaning skeletons with dermestid beetles—two refinements in the method. *Curator* 17:290–98.

Stalling, D. T. 1990. *Microtus ochrogaster. Mamm. Species* 355:1–9.

Streubel, D. P., and J. P. Fitzgerald. 1978. *Spermophilus tridecemlineatus. Mamm. Species* 103:1–5.

Strong, W. 1989. Ecoclimatic regions of Canada, first approximation. *Canadian Widl. Serv., Sustainable Development Branch, Conservation and Protection. Ecological Land Classification Ser.,* 23.

Sudman, P. D., J. C. Burns, and J. R. Choate. 1986. Gestation and postnatal development of the plains pocket gopher. *Texas J. Sci.* 38:91–94.

Svendsen, G. E. 1989. Pair formation, duration of pair-bonds, and mate replacement in a population of beavers (*Castor canadensis*). *Can. J. Zool.* 67:336–40.

Swihart, R. K. 1992. Home-range attributes and spatial structure of woodchuck populations. *J. Mamm.* 73:604–18.

Taber, R. D. 1971. Criteria of sex and age. In *Wildlife management techniques,* ed. R. H. Giles, Jr., 325–402. 3d ed. Washington, D. C.: The Wildlife Society.

Taber, R. D., and I. McT. Cowan. 1971. Capturing and marking wild animals. In *Wildlife management techniques*, ed. R. H. Giles, Jr., 277–318. 3d ed. Washington, D. C.: The Wildlife Society.

Thomas, J. W., and D. Toweill. 1982. *Elk of North America: Ecology and management*. Harrisburg, Penn.: Stackpole Books.

Thomson, C. E. 1982. *Myotis sodalis. Mamm. Species* 163:1–5.

Thorne, D. H., and D. C. Andersen. 1990. Long-term soil-disturbance pattern by a pocket gopher, *Geomys bursarius. J. Mamm.* 71:84–89.

Tidemann, C. R., and D. P. Woodside. 1978. A collapsible bat trap and a comparison of results obtained with the trap and with mist-nets. *Austral. Wildl. Res.* 5:355–62.

Tumlinson, R. 1987. *Felis lynx. Mamm. Species* 269:1–8.

Turner, B. N., S. L. Iverson, and K. L. Severson. 1976. Postnatal growth and development of captive Franklin's ground squirrels (*Spermophilus franklinii*). *Amer. Midl. Nat.* 95:93–102.

Tuttle, M. D. 1974. An improved trap for bats. *J. Mamm.* 55:475–77.

van Zyll de Jong, C. G. 1975. The distribution and abundance of the wolverine (*Gulo gulo*) in Canada. *Can. Field-Nat.* 89:431–37.

———. 1979. Distribution and systematic relationships of long-eared *Myotis* in western Canada. *Can. J. Zool.* 57:987–94.

———. 1983. *Handbook of Canadian mammals. Vol. 1. Marsupials and insectivores*. Ottawa: Natl. Mus. Canada.

———. 1985. *Handbook of Canadian mammals. Vol. 2. Bats*. Ottawa: Natl. Mus. Canada.

Vaughan, T. A. 1986. *Mammalogy*. 3d ed. Philadelphia: Saunders College Publ.

Veatch, J. O. 1953. *Soils and land of Michigan*. East Lansing: Michigan State College Press.

Wade-Smith, J., and B. J. Verts. 1982. *Mephitis mephitis. Mamm. Species* 173:1–7.

Watkins, L. C. 1972. *Nycticeius humeralis. Mamm. Species* 23:1–4.

Webster, W. D., and J. K. Jones, Jr. 1982. *Reithrodontomys megalotis. Mamm. Species* 167:1–5.

Wells-Gosling, N. 1985. *Flying squirrels: Gliders in the dark*. Washington, D. C.: Smithsonian Inst. Press.

Wells-Gosling, N., and L. R. Heaney. 1984. *Glaucomys sabrinus. Mamm. Species* 229:1–8.

Whitaker, J. O., Jr. 1972. *Zapus hudsonius. Mamm. Species* 11:1–7.

———. 1974. *Cryptotis parva. Mamm. Species* 43:1–8.

Whitaker, J. O., Jr., and P. Clem. 1992. Food of the evening bat *Nycticeius humeralis* from Indiana. *Amer. Midl. Nat.* 127:211–14.

Whitaker, J. O., Jr., and S. L. Gummer. 1992. Hibernation of the big brown bat, *Eptesicus fuscus*, in buildings. *J. Mamm.* 73:312–16.

Whitaker, J. O., Jr., and R. E. Wrigley. 1972. *Napaeozapus insignis. Mamm. Species* 14:1–6.

Wilkinson, G. S. 1992a. Communal nursing in the evening bat, *Nycticeius humeralis. Behav. Ecol. Sociobiol.* 31:225–35.

———. 1992b. Information transfer at evening bat colonies. *Anim. Behav.* 44:501–18.

Williams, D. F., and S. E. Braun. 1983. Comparison of pitfall and conventional traps for sampling small mammal populations. *J. Wildl. Manage.* 47:841–45.

Willner, G. R., G. A. Feldhamer, E. E. Zucker, and J. A. Chapman. 1980. *Ondatra zibethicus. Mamm. Species* 141:1–8.

Wilson, D. E. 1982. Wolverine. In *Wild mammals of North America*, ed. J. A. Chapman and G. A. Feldhamer, 644–52. Baltimore: Johns Hopkins Univ. Press.

Wilson, D. E., and D. M. Reeder, eds. 1993. *Mammal species of the world: A taxonomic and geographic reference.* 2d ed. Washington, D. C.: Smithsonian Inst. Press.

Woods, C. A. 1973. *Erethizon dorsatum. Mamm. Species* 29:1–6.

Woolf, A., D. R. Shoemaker, and M. Cooper. 1993. Evidence of tularemia regulating a semi-isolated cottontail rabbit population. *J. Wildl. Manage.* 57:144–57.

Yates, T. L., and D. J. Schmidly. 1975. Karyotype of the eastern mole (*Scalopus aquaticus*), with comments on the karyology of the family Talpidae. *J. Mamm.* 56:902–5.

———. 1978. *Scalopus aquaticus. Mamm. Species* 105:1–4.

Index

Ground squirrel, 281
  Franklin's, 12, **118–20,** 311, 329
  thirteen-lined, 12, **120–23,** 311, 328
  See also Chipmunk; Woodchuck
  Gulo gulo, 10, 222, **273–74,** 318

Hair, 2
Hantavirus, 280
Hare, **91–94,** 304, 309, 323, 327
  European, 13, **102–4,** 309, 327
  snowshoe, **99–102,** 309, 327
  See also Jackrabbit
Harvest mouse, western, 12, 147, **148–50,**
  189, 314, 330
Heat. *See* Estrus
Hibernation
  bats, of, 6, 62–63, 65, 66, 69, 73, 76, 77,
    80, 84, 87, 90
  dispersal to islands, and, 7
  jumping mice, of, 189, 192, 193–94
  squirrels, of, 117, 120, 122
  See also Carnivorean lethargy; Torpor,
    daily
Hopping locomotion, 189, 190, 192, 193
Horns, 258–59
Horse, 304, 322
Humans, impact on mammals, 12–13

Induced ovulation, 98, 122
Insectivora, **24,** 304, 305, 323, 324
Insectivores, **24,** 304, 305, 323, 324
Islands, species diversity on, 7–8

Jackrabbit, white-tailed, 13, **105–7,** 309,
  327
Jumping mouse, 7
  family, **189,** 312, 328
  meadow, 190, **192–95,** 312, 328
  woodland, 148, 161, **189–92,** 312, 328

Kettle hole, 8

Label, specimen, 287–88, 293, 296
Lagomorpha, **91–92,** 304, 309, 323, 327
Lasionycteris noctivagans, **80–82,** 308, 326
Lasiurus
  borealis, **74–77,** 308, 325
  cinereus, 74, **77–79,** 308, 325
Lemming, bog. *See* Bog lemming
Leporidae, **93–94,** 323, 327

Lepus
  americanus, **99–102,** 309, 327
  europaeus, 13, **102–4,** 309, 327
  townsendii, 13, **105–7,** 309, 327
Lichens as food, 272
Lontra canadensis. See Lutra canadensis
Lutra canadensis, **247–50,** 315, 318, 332
Lyme disease, 280
Lynx
  canadensis, 10, 11, 12, **252–54,** 315,
    331
  lynx, 252
  rufus, 10, **254–57,** 315, 331
Lynx, Canadian, 10, 11, 12, **252–54,** 315,
  331

Mammals
  characteristics of, 2–3
  evolution of, 1
  largest, 1
  smallest, 1, 25
Mammary glands, 2
Mange, sarcoptic, 211
Maple sugar, 131
Marmota monax, 7, **115–18,** 311, 329
Marsupials, **17–18**
Marten
  American, 11, 12, **223–26,** 318, 332
  pine. *See* Marten, American
Martes
  americana, 11, 12, **223–26,** 318, 332
  pennanti, 11, 12, **226–28,** 318, 333
Measurements, standard, 288
Meningeal worm, 264, 269, 272
Mephitis mephitis, 2, **244–47,** 280, 319,
  332
Metatheria, **17–18**
Michigan Basin, 4–6
Microsorex hoyi. See Sorex hoyi
Microtinae, 147
Microtus
  chrotorrhinus, 33, **165–67,** 314, 330
  ochrogaster, 12, **167–70,** 314, 330
  pennsylvanicus, **170–72,** 165, 167, 169,
    314, 330
  pinetorum, 11, **173–75,** 314, 329
Milk, 2, 68, 84
Mines, gating of, 71
Mink, 8, **236–39,** 319, 332
Mist net, 285–86

centimeters

0 1 2 3 4 5 6 7 8 9 10 11 12 13 14 15